Birkhäuser

Frontiers in Mathematics

Advisory Editors

Laurent Saloff-Coste, Cornell University, Ithaca, NY, USA

Igor Shparlinski, The University of New South Wales, Sydney, NSW, Australia

Wolfgang Sprößig, TU Bergakademie Freiberg, Freiberg, Germany

This series is designed to be a repository for up-to-date research results which have been prepared for a wider audience. Graduates and postgraduates as well as scientists will benefit from the latest developments at the research frontiers in mathematics and at the "frontiers" between mathematics and other fields like computer science, physics, biology, economics, finance, etc. All volumes are online available at SpringerLink.

Yves Félix • Steve Halperin

Lie Models for Spaces

A New Approach to Rational Homotopy

Yves Félix
Mathematics
Catholic University of Louvain-la-Neuve
Louvain-la-neuve, Belgium

Steve Halperin
Mathematics
University of Maryland, College Park
College Park, MD, USA

ISSN 1660-8046   ISSN 1660-8054 (electronic)
Frontiers in Mathematics
ISBN 978-3-032-15356-2 ISBN 978-3-032-15357-9 (eBook)
https://doi.org/10.1007/978-3-032-15357-9

Mathematics Subject Classification: 55P62, 17B55, 20F14, 55P60

This book is published under the imprint Birkhäuser, www.birkhauser-science.com by the registered company Springer Nature Switzerland AG
The registered company address is: Gewerbestrasse 11, 6330 Cham, Switzerland

If disposing of this product, please recycle the paper.

Preface

Why a new book on Rational Homotopy Theory? In fact, Rational Homotopy has been a success story in algebraic topology for 50 years and is always a theory in building. The first two contributions to the subject, attributed to Quillen and Sullivan [72, 77], reflected the original two meanings of computing homology and cohomology at the start of the twentieth century: simplicial homology of a polyhedron and de Rham cohomology of a manifold. The basic object in the theory is the Sullivan minimal model $(\wedge V, d)$ of a space X. This object, unique up to isomorphism, is a commutative differential graded algebra (cdga) that contains all the rational homotopy information of X. In particular, there is a natural isomorphism of graded algebras $H^*(\wedge V, d) \cong H^*(X; \mathbb{Q})$, and when X is simply connected with finite Betti numbers, then for any $r \geq 2$, $V^r \cong \mathrm{Hom}(\pi_r(X), \mathbb{Q})$.

Now, a geometric realization functor associates with $(\wedge V, d)$ a space $\langle \wedge V, d \rangle$, called the rationalization of X, denoted $X_{\mathbb{Q}}$, and a natural map

$$\varphi_X : X \to X_{\mathbb{Q}}.$$

Once again when X is simply connected with finite Betti numbers then φ_X is a rational homotopy equivalence, $\pi_*(X_{\mathbb{Q}}) \cong \pi_*(X) \otimes \mathbb{Q}$.

Rational Homotopy Theory has been developed in steps. The basic constructions and the induced homotopy equivalence of the homotopy category of rational simply connected spaces with finite Betti numbers with a homotopy category of cdg algebras constituted the first step of the theory. This material is described in Félix et al. [33] and Bousfield and Guggenheim [11], and in many surveys ([46, 52, 81],...)

The second step corresponds to the appearance of a new invariant, the rational Lusternik-Schnirelmann category of a space. The study of this invariant was the main ingredient of many results concerning the rational homotopy of spaces, such as the dichotomy elliptic-hyperbolic or the exponential growth of the ranks of homotopy groups in the hyperbolic case. All this material is described in Félix et al. [33] and [36]. At the same time rational homotopy was used in a very powerful way in various areas of differential topology and geometry: the existence of free torus actions, the structure of configuration spaces and of arrangements, the existence of geodesics and of complex structures, mapping spaces, etc. Part of this development can be found in Allday and

Puppe, Félix et al., Cirici, and Berglund and Madsen [1, 7, 20, 21, 35], etc. Note that the theory is enriched regularly with nice new results (see for instance Milivojevic et al. [66], or Libgober [61]). Also, over time some problems have become better understood, with the solution (in the positive or negative sense) of some of the questions and conjectures. See for instance Amman [2].

At that point all the theory was confined to the limited world of simply connected spaces with finite Betti numbers. However, the minimal Sullivan model $(\wedge V, d)$ and its geometric realization are defined for all path-connected space, X. In Félix et al. [36] we began the study of rational homotopy for general path-connected spaces. We show that $\text{Hom}(V^1, \mathbb{Q})$ has a natural Lie algebra structure isomorphic, via the Baker-Campbell-Hausdorff formulas, to the Malcev completion of $\pi_1(X)$. We also prove that for all path-connected spaces, $\text{Hom}(V^r, \mathbb{Q}) \cong \pi_r(X_{\mathbb{Q}})$ for $r \geq 2$.

The main question consists in knowing when $\varphi_X : X \to X_{\mathbb{Q}}$ is a rational homotopy equivalence. When X has the homotopy type of a finite-type CW complex, φ_X can be identified with the Bousfield $\mathbb{Q}$-completion [10] and the question is to determine the "good spaces" in the sense of Bousfield. In fact, the question remains essentially open even if we give some partial answers that suggest that this happens only if the space X is nilpotent with finite-type homology.

Recall next that, in his fundamental work [72], Quillen associated with a simply connected space X a differential graded Lie (dgl) algebra also containing all the rational information of X. Then, based on an idea of Lawrence and Sullivan [56], an extension of the Quillen-Lie model has been developed in Lazarev and Markl, [58] and in Buijs et al. [15] in the form of complete Lie algebras. More precisely, a complete Lie algebra $\mathcal{L}_X$ is associated with any space X. When X is path connected with finite Betti numbers then $H_0(\mathcal{L}_X)$ is the Malcev completion of $\pi_1(X)$. Moreover, a completed version of the cochain construction enables the Sullivan minimal model to be recovered from $\mathcal{L}_X$.

This book is a next step in Rational Homotopy Theory. With each path-connected space X, we associate a dgl algebra of a special form that we call an *enriched dgl, L_X*. This dgl algebra is closely related to the Sullivan minimal model and coincides with $\mathcal{L}_X$ for spaces with finite-type Betti numbers (they can be different otherwise). The enriched dgl algebras form a new category of algebraic objects and in order to avoid the multiple "pitfalls" we present them in detail with properties and results in the first chapter.

Enriched Lie models are a new and useful way to study the rational homotopy of a space. For instance, when the Sullivan minimal model $(\wedge V, d)$ is quasi-isomorphic to a cdga of the form $(R, 0)$ then its homotopy Lie algebra is a "profree Lie algebra," the enriched analog of a free Lie algebra.

Enriched Lie models have the advantage of being very convenient for the study of cofibrations. An important part of this text is devoted to the application of these new Lie models to extend to the nonsimply connected world the theory of inert maps. We prove that a cell addition $\iota : X \to Y = X \cup e^n$ is rationally inert if the rational homotopy Lie algebra of the homotopy fiber of the rationalization of ι is a profree Lie algebra $\overline{\mathbb{L}}(V)$, where V is isomorphic to a suspension of the complete universal enveloping algebra of $\pi_*(Y_{\mathbb{Q}})$.

With these models we now have a good description of the rational homotopy type of a space. The last section of this book contains applications. Rational Homotopy Theory is research in progress and this book must be thought of as a step in our understanding of the rational homotopy of spaces. We hope that it will be followed by other steps, either with a theoretical aspect or composed of new topological and geometric applications.

Louvain-la-Neuve, Belgium   Yves Félix
College Park, MD, USA   Steve Halperin

Introduction

This book is centered on the construction of a set of convenient new Lie models for path-connected spaces. They are called enriched Lie models and appear as important companions of the traditional Sullivan models. For example, Sullivan models naturally reflect properties of fibrations, whereas enriched Lie models naturally reflect properties of cofibrations and CW complexes. This introduction extends to nonsimply connected spaces the classical relation and Koszul duality between the Sullivan [77] and the Quillen rational homotopy theory [72].

Rational Homotopy Theory is based on a simplicial commutative differential graded algebra (cdga) $A_{PL}(\Delta^n)$ and a Quillen adjunction

$$\langle \cdot \rangle : \qquad \mathrm{cdga} \;\rightleftarrows\; \mathrm{sSet}^{\mathrm{op}} \qquad : A_{PL}$$

First of all, sSet denotes the category of pointed simplicial sets, and A_{PL} is a functor that associates the cdga $\mathrm{Hom}_{sSet}(X, A_{PL}(\Delta^*))$ with a simplicial set X. Second, $\langle \cdot \rangle$ is a functor, called *Sullivan geometric realization*, and defined by

$$\langle A, d \rangle = \mathrm{Hom}_{cdga}((A, d), A_{PL}(\Delta^n)).$$

This adjunction induces the classical Quillen equivalence between the homotopy categories of rational reduced simplicial sets with finite-type homology and that of finite-type 1-connected cdg algebras.

The main advantage of the adjunction is the existence of minimal Sullivan algebras. For recall, a minimal Sullivan algebra is a cdga, $(\wedge V, d)$, in which:

(i) $V = V^{\geq 1}$,

(ii) $\wedge V$ is the free graded commutative algebra generated by V,

(iii) $d : V \to \wedge^{\geq 2} V$, and d satisfies a "Sullivan condition": V is the increasing union of a sequence of subspaces $V = \cup_{n \geq 0} V(n)$ such that $d(V(n)) \subset \wedge V(n-1)$.

The component, $d_1 : V \to \wedge^2 V$ of d, defines by duality a graded Lie bracket in

$$L_V := s^{-1}\mathrm{Hom}(V; \mathbb{Q})$$

(s^{-1} is inverse suspension) called the homotopy Lie algebra of $(\wedge V, d)$.

Central to the Sullivan approach [77] is that any cdga, (A, d), with $H^0(A) = \mathbb{Q}$ admits a quasi-isomorphism $(\wedge V, d) \xrightarrow{\simeq} (A, d)$ from a minimal Sullivan algebra (its minimal Sullivan model.) The *minimal Sullivan model of a space* X is then the minimal model of $A_{PL}(X)$, $(\wedge V, d) \xrightarrow{\simeq} A_{PL}(X)$, determined up to isomorphism by the homotopy type of X.

The *Sullivan rationalization* of X is the map,

$$X \to X_{\mathbb{Q}} = \langle \wedge V \rangle,$$

adjoint to the quasi-isomorphism $\wedge V \xrightarrow{\simeq} A_{PL}(X)$. In particular [77], there is a bijection of graded Lie algebras

$$sL_V \xrightarrow{\cong} \pi_*(X_{\mathbb{Q}}).$$

In the world of simply connected spaces with finite-type rational homology the interplay between cdga and sSet is the best that we could expect. In particular, $\pi_*(X_{\mathbb{Q}}) \cong \pi_*(X) \otimes \mathbb{Q}$ and $H_*(X_{\mathbb{Q}}) = H_*(X; \mathbb{Q})$.

This is not the case in the wide world of nonsimply connected or nonfinite CW complexes. One of our purposes consists in specifying which result remains true and what we can say in general. In particular, the properties of the rationalization map are essentially unknown. Partial answers are given in Section 10 concerning the determination of the Sullivan rational spaces, i.e., of those spaces for which the rationalization map is a homotopy equivalence. For instance, in Proposition 10.3 we give a characterization of simply connected Sullivan rational spaces.

Now we introduce enriched dgl algebras.

Definition *An enriched dgl is a dgl* $(L = L_{\geq 0}, \partial)$ *together with an inverse system of surjections*

$$\rho_\alpha : L \to L_\alpha$$

onto finite-dimensional nilpotent dgl algebras, such that

$$L \xrightarrow{\cong} \varprojlim_\alpha L_\alpha.$$

In particular, this equips (L, ∂) with a topology, and we can (and do) always complete subspaces and sub-Lie algebras with respect to that topology.

Analogous to the role of Sullivan algebras in the category of cdg algebras we introduce the *profree Lie algebras* $\overline{\mathbb{L}}_T$, and the *profree* dgl algebras $(\overline{\mathbb{L}}_T, \partial)$. Here, $\overline{\mathbb{L}}_T$ is the completion of a free graded Lie algebra, $\mathbb{L}_T$, on a closed subspace T. More precisely when T is a finite-type graded vector space, then $\overline{\mathbb{L}}_T$ is the inverse limit of the quotients $\mathbb{L}_T / \mathbb{L}_T^n$ of $\mathbb{L}_T$ by its central series. In the general case, $T = \varprojlim_\alpha T_\alpha$ is the inverse limit of a system of finite-dimensional vector spaces T_α and $\overline{\mathbb{L}}_T = \varprojlim_\alpha \overline{\mathbb{L}}_{T_\alpha}$.

A profree dgl $(\overline{\mathbb{L}}_T, \partial)$ is called minimal if $\partial : T \to [\overline{\mathbb{L}}_T, \overline{\mathbb{L}}_T]$. Completing the analogy we show that any enriched dgl, (L, ∂), admits a quasi-isomorphism

$$(\overline{\mathbb{L}}_T, \partial) \xrightarrow{\cong} (L, \partial)$$

from a minimal profree dgl, and that this determines $(\overline{\mathbb{L}}_T, \partial)$ up to isomorphism (Proposition 13.3).

If (L, ∂) is an enriched dgl algebra, its associated Sullivan model is the generalized cochain algebra $\widehat{C}(L, \partial)$, denoted $(\wedge V_L, d)$, and defined as the direct limit

$$\widehat{C}(L, \partial) := \varinjlim_\alpha C^*(L_\alpha, \partial),$$

where C^* is the usual cochain algebra functor. In particular, $\wedge V_L$ is a semi-quadratic Sullivan algebra, which means that $d = d_0 + d_1, d_0 : V_L \to V_L$ and $d_1 : V_L \to \wedge^2 V_L$.

Theorem *The enriched dgl algebras form a cofibrantly generated model category denoted Enrich and the generalized cochain construction $\widehat{C}$ induces a Quillen equivalence of homotopy categories*

$$cdga \longrightarrow Enrich$$

This construction makes enriched dgl algebras the Koszul duals of the cdg algebras.

In particular, the construction associates a minimal profree dgl algebra $(\overline{\mathbb{L}}_T, \partial)$ with a connected space X (via $A_{PL}(X)$) such that

$$H(\overline{\mathbb{L}}_T, \partial) = \pi_*(X_{\mathbb{Q}}) \quad \text{and} \quad \mathbb{Q} \oplus sT \cong (H^*(X; \mathbb{Q}))^\vee.$$

As in the case of minimal Sullivan models, $(\overline{\mathbb{L}}_T, \partial)$ is determined up to isomorphism by the homotopy type of X. The interplay between these two equivalent linked models for a connected space is the basis for the properties and examples that we construct.

For example, whereas the classical Sullivan models provide simple descriptions of the fiber of a fibration, profree dgl algebra models provide simple descriptions of the cofiber of a cofibration. In each case the appropriate choice of model can enable a computational approach as well as the construction of interesting examples.

More precisely, if $\gamma : (\overline{\mathbb{L}}_T, 0) \to (L, \partial)$ is an enriched dgl algebra representative of a continuous map

$$g : \vee_\alpha S^{n_\alpha} \to X,$$

then an enriched dgl algebra model of $X \cup_g \left(\vee_\alpha D^{n_\alpha+1} \right)$ is given by the free Lie product

$$(L \,\widehat{\amalg}\, \overline{\mathbb{L}}_{sT}, \partial + \partial_\gamma)$$

where $\partial_\gamma(L) = 0$ and $\partial_\gamma(sx) = \gamma(x)$. This construction extends to provide a dgl algebra model of any CW complex.

This interplay between Sullivan models and enriched dgl algebra models is also a powerful tool for understanding the following basic question: when does a cell attachment

$$\iota_X : X \to X \cup_g \vee_\alpha D^{n_\alpha+1}$$

lead to a surjection

$$\pi_*(\iota_X)_{\mathbb{Q}} : \pi_*(X_{\mathbb{Q}}) \to \pi_* \left(X \cup_g \vee_\alpha D^{n_\alpha+1} \right)_{\mathbb{Q}}.$$

In this case we say that g is *rationally inert*. We provide specific characterizations of rational inertness, one specific to Sullivan models and another in terms of enriched dgl algebra models. For instance, extending results obtained in [41] for simply connected spaces, we show in Part IV that

- The attachment of a single cell to a wedge of odd spheres is rationally inert if the attaching map is not rationally trivial, and
- If Y is an n-dimensional finite CW complex, and its cohomology algebra $H^*(Y; \mathbb{Q})$ satisfies Poincaré duality and has at least two generators, then Y is obtained from its $(n-1)$-skeleton by a rationally inert attachment.

Rational homotopy theory begins with three fundamental contributions: Quillen's [72] (for simply connected spaces), Bousfield and Kan's [12], and Sullivan's [77]. We note that for CW complexes of finite type the latter two are equivalent [11]. We cannot forget the work of Chen [17], who, in 1977, associated a dgl algebra of the form $(\overline{\mathbb{L}}(V), d)$ where V is the desuspension of the reduced homology of V, with a manifold or a simplicial set X. When X is simply connected, Chen showed that its enveloping algebra is isomorphic to the loop space homology of X as a Hopf algebra (see also Hain [39]).

More recently, Lazarev and Markl [58], and Hinich [50], have used enriched dgl algebras (without calling them enriched) to construct an equivalence of homotopy categories between cdg algebras and enriched dgl algebras in the more general situation of non-necessarily connected spaces [58, Proposition 9.16]. Here we work only with connected spaces and focus on applications to topology of the correspondence between minimal Sullivan models and profree models.

A different approach in this vein is attributed to Buijs et al. [15]. There, the authors associate a pronilpotent dgl algebra M_X with each finite simplicial set X. For finite simplicial spaces, the models are quasi-isomorphic. More details are given in Sect. 13.3.

The text is divided into four parts. Part I contains the basic theory of enriched Lie algebras and associated quadratic Sullivan algebras. Theorem 6.1 (Chap. 6) states that an enriched Lie algebra is profree if and only if its associated Sullivan algebra $(\wedge V, d)$ is quasi-isomorphic to a cdga of the form $(\mathbb{Q} \oplus R, 0)$ with $R \cdot R = 0$. Minimal Sullivan algebras and Sullivan rationalizations are then described in Part II.

Part III explores the relations among enriched dgl algebra models, Sullivan models, and topological spaces. The connection between enriched dgl algebras and cdg algebras is realized using a generalization of the cochain algebra functor. This part contains all the theory needed for the computation of explicit examples and for developing interesting applications.

Part IV concerns inert cell attachments and their applications. For instance, we prove that any nonzero element is inert in $\overline{\mathbb{L}}_T$, when $T = T_{even}$. We also give a positive answer to the rational version of a Whitehead problem: suppose that X is a connected CW complex and that the addition of two-dimensional cells to X gives an aspherical space, does that imply that X is aspherical?

Conventions (see [36, Chap 1] for details)

- A *topological space* is either a simplicial set or a CW complex, and the two homotopy categories are identified by the singular simplex and the Steenrod realization functors [64]. In either case a connected space is path connected.
- By a sphere we mean a *connected sphere S^n, $n \geq 1$*.
- If X is a topological space, then for simplicity, $H(X) = H^*(X; \mathbb{Q})$ denotes the rational singular cohomology algebra of X. We use the standard notation for homology and for singular cohomology with other coefficients.
- Graded objects are families $Q = \{Q^n\}_{n \in \mathbb{Z}}$, where some of the Q^n may be equal to 0. We adopt the convention $\{Q^n\}_{n \in \mathbb{Z}} = \{Q_{-n}\}_{n \in \mathbb{Z}}$. If $x \in Q$, then $(-1)^{deg\, x} = (-1)^n$ whether $x \in Q^n$ or $x \in Q_n$.
- If $Q = \{Q^n\}$ is a graded object, then its suspension sQ is the graded object defined by $(sQ)_n = Q_{n-1}$.
- Because the notation $(\ )^*$ is used everywhere, and because dual vector spaces appear here so frequently, we simplify the notation $V^{\vee} = \mathrm{Hom}(V, \mathbb{Q})$ and write

$$(V^{\vee})_n = \mathrm{Hom}(V^n, \mathbb{Q})$$

for any graded vector space V.

- A *graded Lie algebra* is a graded vector space L together with a linear map of degree 0, $[\,,\,] : L \otimes L \to L$, satisfying

$$[x, y] + (-1)^{deg\,x\,deg\,y}[y, x] = 0 \quad \text{and} \quad [x, [y, z]] = [[x, y], z] + (-1)^{deg\,x\,deg\,y}[y, [x, z]].$$

- A *differential graded Lie algebra* (dgl) is a graded Lie algebra L together with a differential ∂ of degree -1, which is also a derivation:

$$\partial([x, y]) = [\partial x, y] + (-1)^{deg\,x}[x, \partial y].$$

- A *commutative graded algebra*, (A, d) is a graded algebra $A = \{A^n\}_{n \geq 0}$ satisfying $ab = (-1)^{deg\,a\,deg\,b}\,ba$ and $H^0(A) = \mathbb{Q}$.
- A *commutative differential graded algebra* (cdga) is an augmented commutative graded algebra A with a derivation of degree $+1$ satisfying $d(ab) = d(a)b + (-1)^{deg\,a}ad(b)$. An *augmentation* is a surjective morphism $\varepsilon_A : (A, d) \to (\mathbb{Q}, 0)$.
- A *Sullivan path object* for an augmented cdga, $\varepsilon_A : A \to \mathbb{Q}$, is a decomposition of the diagonal map $(id, id) : A \to A \times A$ as

$$A \to A \otimes \wedge(S, dS) \xrightarrow{\rho} A \times A$$

in which $S = S^{\geq 0}$ and $d : S \xrightarrow{\cong} dS$ and ρ is surjective. In particular, the morphism $(\varepsilon_0, \varepsilon_1) : A \otimes \wedge(t, dt) \to A \times A$ defined by $\varepsilon_0(t) = 0$, $\varepsilon_1(t) = 1$, is a Sullivan path object. It is straightforward to check that two Sullivan path objects for A are connected by a quasi-isomorphism, φ, that is, the identity on A and satisfies $\rho' \circ \varphi = \rho$.

Two augmentation-preserving morphisms $\varphi_0, \varphi_1 : \wedge V \to A$ are *homotopic* ($\varphi_0 \sim \varphi_1$) if $(\varphi_0, \varphi_1) : \wedge V \to A \times A$ lifts through ρ to a morphism (called the *homotopy*),

$$\varphi : \wedge V \to A \otimes \wedge(t, dt),$$

satisfying $\varphi(V) \subset A^{\geq 1} \otimes \wedge(t, dt)$. This condition is independent of the choice of path object, and homotopy is an equivalence relation.

- A *Sullivan algebra* is a cdga $(\wedge V, d)$ in which

  (i) $\wedge V$ is a free graded commutative algebra and $V = V^{\geq 1}$. We denote by $\wedge^p V$ the space spanned by the products of length p of elements in V.
  (ii) $V = \cup_{n \geq 0} V_n$, where $V_0 = V \cap \ker d$ and $V_{n+1} = V \cap d^{-1}(\wedge V_n)$.

- A Sullivan algebra $(\wedge V, d)$ is *minimal* if $d : V \to \wedge^{\geq 2} V$, and is *quadratic* if $d : V \to \wedge^2 V$; moreover, a quasi-isomorphism between minimal Sullivan algebras is an isomorphism.
- A (minimal) Λ-*extension* is a sequence of cdga morphisms

$$A \xrightarrow{\ \eta\ } A \otimes \wedge Z \xrightarrow{\ \rho\ } \wedge Z$$

in which: $\eta(a) = a \otimes 1$, $\rho = \varepsilon_A \otimes id$ with ε_A an augmentation, and the quotient $\wedge Z$ is a (minimal) Sullivan algebra.

- Each Sullivan algebra $\wedge V$ has an *acyclic closure* [36, Chap 3]. This is a Λ-extension $\wedge V \to \wedge V \otimes \wedge U \to \wedge U$ in which

$$du = p(u) + \Phi, \qquad u \in U,$$

$p : U^k \xrightarrow{\cong} V^{k+1}$, $k \geq 0$, and $\Phi \in \wedge^{\geq 2}(U \oplus V)$.
- In a category $\mathcal{C}$, if $X, Y \in \mathcal{C}$, $\mathcal{C}(X, Y)$ denotes the set of morphisms from X to Y.

Contents

Part I

Enriched Lie Algebras

Enriched and Pre-enriched Lie Algebras

By a *pre-enriched Lie algebra* we mean a graded Lie algebra, $L = L_{\geq 0}$, equipped with a family $\mathcal{J} = \{\rho_\alpha : L \to L_\alpha\}$ of surjective morphisms satisfying the following conditions:

(i) Each L_α is a finite dimensional nilpotent Lie algebra.
(ii) The index set, $\{\alpha\}$, is a directed set under the partial order given by

$$\alpha \geq \beta \iff \ker \rho_\alpha \subset \ker \rho_\beta;$$

that is, for each α, β there is a γ such that $\ker \rho_\gamma \subset \ker \rho_\alpha \cap \ker \rho_\beta$.
(iii) $\cap_\alpha \ker \rho_\alpha = 0$.

Moreover, we shall use $\mathcal{J}$ to denote both the family $\{\rho_\alpha\}$ and the index set $\{\alpha\}$ and denote $\ker \rho_\alpha$ by I_α.

Example In any graded Lie algebra $L = L_{\geq 0}$ the set of all surjective morphisms onto finite dimensional and nilpotent graded Lie algebras forms a directed system as above. If the intersection of the kernels is zero, then this defines a pre-enriched structure.

Definition A *morphism* $\varphi : (E, \mathcal{J}_E) \to (F, \mathcal{J}_F)$ of pre-enriched Lie algebras is a morphism of graded Lie algebras such that for each $\alpha \in \mathcal{J}_F$ there is a $\beta \in \mathcal{J}_E$ such that $\varphi : \ker \rho_\beta \to \ker \rho_\alpha$.

This is equivalent to say that for any $\alpha \in \mathcal{J}_F$ there is some $\beta \in \mathcal{J}_E$ such that $\rho_\alpha \circ \varphi$ factors through some ρ_β

Y. Félix, S. Halperin, *Lie Models for Spaces*, Frontiers in Mathematics,
https://doi.org/10.1007/978-3-032-15357-9_1

$$\begin{array}{ccc} E & \xrightarrow{\varphi} & F \\ \downarrow{\scriptstyle\rho_\beta} & & \downarrow{\scriptstyle\rho_\alpha} \\ E_\beta & \dashrightarrow & F_\alpha. \end{array}$$

1.1   The Completion of a Pre-enriched Lie Algebra

Definition An *enriched Lie algebra* is a pre-enriched Lie algebra $(L, \mathfrak{I}_L)$ for which $L = \varprojlim_\alpha L_\alpha$.

If $(L, \mathfrak{I}_L)$ is a pre-enriched Lie algebra, we define $(\overline{L}, \mathfrak{I}_{\overline{L}})$ by setting $\overline{L} = \varprojlim_\alpha L_\alpha$ and $\mathfrak{I}_{\overline{L}} = \{\overline{\rho_\alpha} : \overline{L} \to L_\alpha\}$, where the $\overline{\rho_\alpha}$ are the projections of the inverse limit.

It follows from the defining properties of pre-enriched Lie algebras that $(\overline{L}, \mathfrak{I}_{\overline{L}})$ is an enriched Lie algebra, that the map $\lambda_L : L \to \overline{L}$ induced by the ρ_α is injective, and that

$$\lambda_L : (L, \mathfrak{I}_L) \to (\overline{L}, \mathfrak{I}_{\overline{L}})$$

is a morphism of pre-enriched Lie algebras. Moreover, almost by definition, we have

$$L \cap \ker \overline{\rho_\alpha} = \ker \rho_\alpha. \tag{1.1}$$

Definition The morphism λ_L is the *completion* of $(L, \mathfrak{I}_L)$.

Examples of Completion
1. Let L be a finitely generated graded Lie algebra, and let

$$L = L^1 \supset \cdots \supset L^k \supset \ldots$$

be its lower central series: L^k is the linear span of iterated commutators of length k. Then each L^k is an ideal, and if $\cap_k L^k = 0$, then the completion $\overline{L}$ is the pronilpotent completion $\overline{L} = \varprojlim_n L/L^n$
2. A Lie algebra may admit different nonequivalent pre-enriched structures. For instance, let $L = \mathbb{Q}[x]$ viewed as an abelian Lie algebra. The family of surjections $\rho_n : L \to L/x^n.L$ then makes L a pre-enriched Lie algebra for which $\overline{L} = \mathbb{Q}[[x]]$.

    Now add to the ρ_n the evaluation of x at 1, $ev : \mathbb{Q}[x] \to \mathbb{Q}$. This gives another pre-enriched structure to $\mathbb{Q}[x]$ and $\overline{L} = \mathbb{Q}[[x]] \oplus \mathbb{Q}$.

Finally, observe that any morphism $\psi : (E, \mathfrak{I}_E) \to (F, \mathfrak{I}_F)$ of pre-enriched Lie algebras extends naturally to a morphism $\overline{\psi}$ between their completions. In fact, set $\mathfrak{I}_{E,\alpha} = \{\beta \mid \ker \rho_\beta \subset \psi^{-1}(\ker \rho_\alpha)\}$. Then ψ induces morphisms

$$\psi_\alpha : \varprojlim_{\beta \in \mathcal{I}_{E,\alpha}} E_\beta \to F_\alpha.$$

Moreover, $\overline{E} = \varprojlim_\alpha \varprojlim_{\beta \in \mathcal{I}_{E,\alpha}} E_\beta$, and ψ extends to the morphism

$$\overline{\psi} = \varprojlim_\alpha \psi_\alpha : \overline{E} \to \overline{F}.$$

It is straightforward to check that $\overline{\psi}$ is the unique morphism extending ψ.

Definition The *closure* of a subspace S in an enriched Lie algebra $(L, \{\rho_\alpha\})$ is the subspace $\overline{S} := \varprojlim_\alpha \rho_\alpha(S)$.

Remarks

(1) It is immediate that if E is a sub-Lie algebra of an enriched Lie algebra $(L, \mathcal{I}_L)$, then its closure $(\overline{E}, \mathcal{I}_{\overline{E}})$ is a sub-enriched Lie algebra of $(\overline{L}, \mathcal{I}_{\overline{L}})$.
(2) If $S \subset L$, then $\rho_\alpha(S) = \rho_\alpha(\overline{S})$.

Next, if $\mathcal{I}$ and $\mathcal{I}'$ are families satisfying the conditions above for L, we say that $\mathcal{I}$ and $\mathcal{I}'$ are *equivalent* ($\mathcal{I} \sim \mathcal{I}'$) if the identities $(L, \mathcal{I}) \to (L, \mathcal{I}')$ and $(L, \mathcal{I}') \to (L, \mathcal{I})$ are morphisms. This is an equivalence relation, and if $\varphi : (F, \mathcal{G}) \to (L, \mathcal{I})$ is a morphism, then $\varphi : (F, \mathcal{G}') \to (L, \mathcal{I}')$ is also a morphism whenever $\mathcal{G} \sim \mathcal{G}'$ and $\mathcal{I} \sim \mathcal{I}'$. In particular, if $\mathcal{I}$ and $\mathcal{I}'$ are equivalent families, then the identity induces an isomorphism

$$\varprojlim_{\alpha \in \mathcal{I}} L_\alpha \xrightarrow{\;\cong\;} \varprojlim_{\beta \in \mathcal{I}'} L_\beta.$$

Thus the completion, $\lambda_L : L \to \overline{L}$, is independent of the choice of $\mathcal{I}$ in its equivalence class.

Examples

(1) Let $(L, \{\rho_\alpha\})$ be a pre-enriched Lie algebra, and let $\{\rho_\beta\}$ be a subfamily such that for each α there is a β with $\ker \rho_\beta \subset \ker \rho_\alpha$. Then (ρ_α) and (ρ_β) are equivalent families.
(2) *Full pre-enriched Lie algebras.* A pre-enriched Lie algebra $(L, \mathcal{I})$ is *full* if whenever an ideal I contains some $\ker \rho_\beta$, then $I = \ker \rho_\alpha$ for some α. If $(L, \mathcal{J})$ is any pre-enriched Lie algebra, then $\mathcal{J}$ extends to a unique full enriched family $\mathcal{I}$ consisting of all the ideals containing some $\ker \rho_\beta$. Evidently $\mathcal{I} \sim \mathcal{J}$. Moreover if $\mathcal{I}$ and $\mathcal{J}$ are equivalent full pre-enriched structures, then they coincide.
(3) A finite dimensional nilpotent Lie algebra L admits up to equivalence only one enriched structure that given by $id : L \to L$.

The following lemma follows directly from the definition of morphisms.

Lemma 1.1 *Let $(L, \{\rho_\alpha\})$ be an enriched Lie algebra. Then:*

(1) Each morphism $\rho : L \to L'$ onto a finite dimensional nilpotent Lie algebra L' factors through some ρ_α.

(2) L is the inverse limit of the surjective morphisms $\rho_\beta : L \to L_\beta$ onto finite dimensional nilpotent Lie algebras.

A key aspect of an enriched Lie algebra $(L, \mathfrak{I})$ is that $\mathfrak{I}$ makes the Lie algebra accessible to finite dimensional arguments, even when L is not a graded vector space of finite type. In particular, enriched Lie algebras behave well with respect to inverse limits, as follows from the following special case of a Theorem in [9]:

Proposition 1.1 (Bourbaki) *Suppose $0 \to A_\alpha \to B_\alpha \to C_\alpha \to 0$ is a directed system of short exact sequences of graded vector spaces of finite type. Then*

$$0 \to \varprojlim_\alpha A_\alpha \to \varprojlim_\alpha B_\alpha \to \varprojlim_\alpha C_\alpha \to 0 \tag{1.2}$$

is also short exact. In particular, for chain complexes of finite type, homology commutes with inverse limits.

Proof A short exact sequence

$$0 \to A_\alpha \to B_\alpha \to C_\alpha \to 0$$

of graded vector spaces of finite type dualizes to a short exact sequence, and an inverse system of such short exact sequences is the dual of the inductive systems of their duals. Since direct limits preserve short exact sequences, it follows that

$$0 \leftarrow \varinjlim_\alpha A_\alpha^\vee \leftarrow \varinjlim_\alpha B_\alpha^\vee \leftarrow \varinjlim_\alpha C_\alpha^\vee \leftarrow 0$$

is short exact. Dualizing again shows that

$$0 \to \varprojlim_\alpha A_\alpha \to \varprojlim_\alpha B_\alpha \to \varprojlim_\alpha C_\alpha \to 0$$

is short exact.

$\square$

The first properties of enriched Lie algebras are given by the following lemma:

Lemma 1.2 *Let L be an enriched Lie algebra. Then:*

(i) Let $I \subset L$ be a closed ideal; then L/I is an enriched Lie algebra.
(ii) A finite sum of closed subspaces is closed.
(iii) If $S \subset L$ has finite type, then $[S, L]$ is closed.

Proof

(i) By Proposition 1.1, we have

$$\overline{L/I} = \varprojlim_{\alpha} L_\alpha/\rho_\alpha(I) = \varprojlim_{\alpha} L_\alpha \,/\, \varprojlim_{\alpha} \rho_\alpha(I) = L/I.$$

(ii) It is sufficient to show that if S and R are closed, then $S + R$ is closed. By definition,

$$\overline{S + R/S} = \varprojlim_{\alpha} \rho_\alpha(S + R)/\rho_\alpha(S) = \varprojlim_{\alpha} \frac{\rho_\alpha(S) + \rho_\alpha(R)}{\rho_\alpha(S)}$$

$$= \varprojlim_{\alpha} \frac{\rho_\alpha(R)}{\rho_\alpha(S) \cap \rho_\alpha(R)}$$

$$= \overline{R}/\varprojlim_{\alpha} \rho_\alpha(S) \cap \rho_\alpha(R).$$

In particular, it follows that the inclusions of $\overline{S}$ and $\overline{R}$ in $\overline{S + R}$ define a surjection $\overline{S} + \overline{R} \to \overline{S + R}$. Since S and R are closed, $S + R \to \overline{S + R}$ is an isomorphism.

(iii) In view of (ii) it is sufficient to prove that for any $x \in L$, $[x, L]$ is closed. Let $C_\alpha = \{y_\alpha \in L_\alpha \,|\, [\rho_\alpha x, y_\alpha] = 0\}$. It is immediate from the definition that $\{C_\alpha\}$ is an inverse system, and we set

$$C = \varprojlim_{\alpha} C_\alpha.$$

Now we show that $C = \{y \in L \,|\, [x, y] = 0\}$. Indeed, if $y \in C$, then for all α,

$$\rho_\alpha[x, y] = [\rho_\alpha x, \rho_\alpha y] \in [\rho_\alpha x, C_\alpha] = 0,$$

and hence $[x, y] = 0$. On the other hand, if $[x, y] = 0$, then $[\rho_\alpha x, \rho_\alpha y] = \rho_\alpha[x, y] = 0$ and so each $\rho_\alpha y \in C_\alpha$. Thus $y \in C$.

Finally, apply (i) to obtain the commutative diagram

$$L/C = \varprojlim_{\alpha} L_\alpha \,/\, \varprojlim_{\alpha} C_\alpha \xrightarrow{\;=\;} \varprojlim_{\alpha} L_\alpha/C_\alpha$$

$$\cong \downarrow [x,-] \qquad\qquad\qquad\qquad \cong \downarrow [\rho_\alpha x,-]$$

$$[x, L] \xrightarrow{\;=\;} \varprojlim_{\alpha} [\rho_\alpha x, L_\alpha]$$

Since $\varprojlim_{\alpha} [\rho_\alpha x, L_\alpha] = \overline{[x, L]}$, (iii) is established. $\qquad\qquad\square$

1.2 Universal Enveloping Algebras

Recall next that the classical completion of the universal enveloping algebra UE of a finite type graded Lie algebra E is the inverse limit

$$\widehat{UE} = \varprojlim_{n} UE/J^n,$$

J^n denoting the nth power of the augmentation ideal. In particular

$$\widehat{UE} = \mathbb{Q} \oplus \widehat{J},$$

and $\widehat{J} := \varprojlim_{n} J/J^n$ is the augmentation ideal for $\widehat{UE}$.

Definition The *completed universal enveloping algebra* $\overline{UL}$ for an enriched Lie algebra $(L, \{\rho_\alpha\})$ is defined by

$$\overline{UL} := \varprojlim_{\alpha} \widehat{UL_\alpha} = \varprojlim_{n,\alpha} UL_\alpha/J_\alpha^n.$$

Passing to inverse limits shows that the inclusions $L_\alpha \to L_\alpha/J_\alpha^n$ define an inclusion

$$\overline{L} \to \overline{UL}.$$

Note also that

$$\overline{UL} = \mathbb{Q} \oplus \overline{J},$$

where $\overline{J} = \varprojlim_{\alpha,n} J_\alpha/J_\alpha^n$ is the augmentation ideal for $\overline{UL}$. It is also immediate that a morphism $\varphi : E \to L$ of enriched Lie algebras extends to a morphism $\overline{U\varphi} : \overline{UE} \to \overline{UL}$.

Finally, in analogy with the filtration of UL by the ideals J^n, we filter $\overline{UL}$ by the ideals

$$\overline{J}^{(n)} := \varprojlim_{k,\alpha} J_\alpha^n / J_\alpha^{k+n} \subset \overline{J} = \overline{J}^{(1)}.$$

Then $\overline{UL}$ is complete with respect to this filtration:

$$\overline{UL} = \varprojlim_n \overline{UL}/\overline{J}^{(n)}. \tag{1.3}$$

In fact, Proposition 1.1 yields

$$\overline{UL} = \varprojlim_{n,\alpha} UL_\alpha / J_\alpha^n = \varprojlim_k \varprojlim_{n,\alpha} (UL_\alpha/J_\alpha^{n+k})/(J_\alpha^n/J_\alpha^{n+k})$$

$$= \varprojlim_n \varprojlim_{k,\alpha} (UL_\alpha/J_\alpha^{n+k})/\varprojlim_{k,\alpha} (J_\alpha^n/J_\alpha^{n+k})$$

$$= \varprojlim_n \overline{UL}/\overline{J}^{(n)}.$$

Example When $L/[L,L]$ has finite type, then $\overline{UL} = \widehat{UL}$.

Remark In analogy with complete enriched Lie algebras, a *complete enriched algebra* is the projective limit of an inductive system of finite dimensional nilpotent graded algebras. The graded commutator makes a complete enriched algebra into a complete enriched graded Lie algebra. The induced functor from complete enriched algebras to complete enriched Lie algebras has a left adjoint functor that associates with a Lie algebra L its universal enveloping algebra $\overline{UL}$.

The structure of the completed universal enveloping algebra of an enriched Lie algebra is also described in [45], where the author describes a generalization of the Poincaré-Birkhoff-Witt theorem.

The (graded) enriched vector spaces are the duals of usual (graded) vector spaces. Since a vector space is the union of its finite dimensional subvector spaces, an enriched vector space is the inverse limit of finite dimensional vector spaces.

2.1   Basic Definitions

Definitions

(1) A *pre-enriched vector space* is a graded vector space T, together with a family $\mathfrak{I}_T$ of surjective linear maps $\rho_\alpha : T \to T_\alpha$ of degree zero, and satisfying:
   (i) Each $\dim T_\alpha < \infty$.
   (ii) The index set $\{\alpha\}$ is a directed set under the partial order given by

$$\alpha \geq \beta \iff \ker \rho_\alpha \subset \ker \rho_\beta.$$

   (iii) $\cap_\alpha \ker \rho_\alpha = 0$.
(2) A linear map $\varphi : (S, \{\rho_\beta\}) \to (T\{\rho'_\alpha\})$ is called *coherent* if for each α there is a β such that the composite $\rho_\alpha \circ \varphi$ factors through ρ'_β.
   A *morphism of pre-enriched vector spaces* is a coherent linear map of degree 0.

Remark In what follows we will define enriched differential graded Lie algebras (L, ∂). The differentials ∂ will be coherent linear maps of degree -1.

Now for pre-enriched vector spaces the following definitions extend those in Chap. 1.

Y. Félix, S. Halperin, *Lie Models for Spaces*, Frontiers in Mathematics,
https://doi.org/10.1007/978-3-032-15357-9_2

Definitions

(1) The *completion* of a pre-enriched vector space $(T, \mathcal{I}_T)$ is the inclusion

$$\lambda_T : (T, \mathcal{I}_T) \to (\overline{T}, \mathcal{I}_{\overline{T}})$$

in which $\overline{T} = \varprojlim_\alpha T_\alpha$ and $\mathcal{I}_{\overline{T}}$ is the set of surjections $\varprojlim_\alpha T_\alpha \to T_\alpha$.

(2) An *enriched vector space* is a complete pre-enriched vector space.

(3) *Sub- and quotient spaces.* Suppose $(T, \mathcal{I}_T)$ is a pre-enriched vector space, and $S \subset T$ is any subspace. The pre-enriched structure in T endows S and T/S with the sub- and quotient pre-enriched structures given, respectively, by

$$\mathcal{I}_S = \{\rho_\alpha : S \to \rho_\alpha(S) \,|\, \alpha \in \mathcal{I}_T\} \quad \text{and} \quad \mathcal{I}_{T/S} = \{\overline{\rho_\alpha} : T/S \to \rho_\alpha T/\rho_\alpha S \,|\, \alpha \in \mathcal{I}_T\}.$$

Remark The enriched graded vector spaces are more known as graded profinite vector spaces and sometimes go under the heading of linearly compact vector spaces.

Examples

(1) The completion $\overline{L}$ of a pre-enriched Lie algebra L and the completion $\overline{UL}$ of its enveloping algebra are examples of completions of pre-enriched vector spaces.

(2) A vector space can have enriched structures with nonisomorphic completions. For instance, let $V = \mathbb{Q}[[x]] = \varprojlim_n \mathbb{Q}[x]/x^n$. Then V is complete for the projections $\rho_n : V \to \mathbb{Q}[x]/x^n$.

Now denote by $\omega = \sum a_n x^n$ a series with $a_n \neq 0$ for all n, and let $S \subset V$ be a direct summand of $\mathbb{Q}\omega$ in V. Finally denote by $\varepsilon : V \to \mathbb{Q}\omega$ the projection with kernel S. The morphisms $\sigma_n : V \to \mathbb{Q}[x]/x^n$ together with $\varepsilon : V \to \mathbb{Q}$ define a new pre-enriched structure on V. Its completion is the injection $id \oplus \varepsilon : \mathbb{Q}[[x]] \to \mathbb{Q}[[x]] \oplus \mathbb{Q}$.

Remarks

(1) A subspace $S \subset T$ is closed if and only if each subspace S_k of degree k is closed.

(2) If $(T, \mathcal{I}_T)$ is an enriched vector space, then the subspaces $\ker \rho_\beta$, $\beta \in \mathcal{I}_T$, are closed. In fact, if $\alpha \geq \beta$, then ρ_β factors over ρ_α to yield $\overline{\rho_\beta} : T_\alpha \to T_\beta$. Thus the remark follows from the commutative diagram

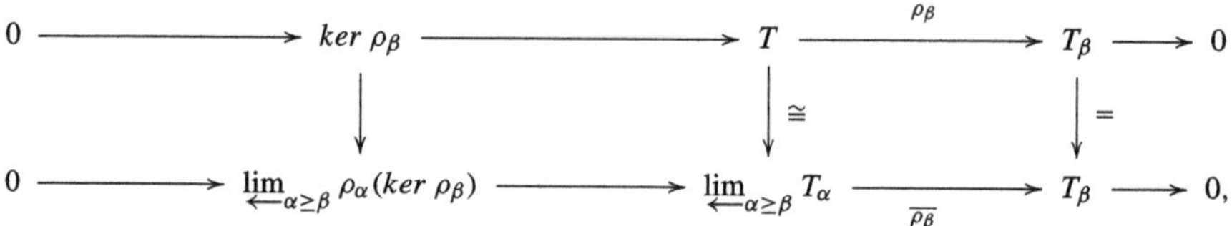

which is row-exact by Proposition 1.1

Example Let $T = \prod_{i=1}^{\infty} V_i$ with the enriched structure given by the projections $\rho_n : T \rightarrow \prod_{i=1}^{n} V_i$. Suppose $V_i = \mathbb{Q}$ for all i, so that T identifies with the set of sequences of rational numbers. Denote by S the set of sequences x_i with $x_i = x_{i+2}$ for all i. Then S is a closed subset. On the other hand denote by S' the set of sequences x_i such that the series $\sum x_i$ is convergent. Then for all n, $\rho_n(S') = \rho_n(T)$ so that $\overline{S'} = T$.

Suppose $(T, \mathfrak{I}_T)$ is an enriched vector space. The advantage of the condition $\dim T_\alpha < \infty$ is that for these spaces $T_\alpha \rightarrow (T_\alpha^\vee)^\vee$ is an isomorphism. Moreover the construction $(\ )^\vee$ converts inductive systems to inverse systems. Therefore, since any graded vector space V is the direct limit, $\varinjlim_\alpha V_\alpha$, of its finite dimensional subspaces, we can associate with V the enriched graded vector space $\varprojlim_\alpha V_\alpha^\vee$.

Proposition 2.1 *The association $V \mapsto V' = \varprojlim_\alpha V_\alpha^\vee$ induces a contravariant isomorphism of categories between the category of graded vector spaces and the category of enriched vector spaces. The duality provides a bijection between the short exact sequences in the two categories, and in both categories such sequences are split.*

Proof The converse construction proceeds as follows. To an enriched graded vector space (V, ρ_α), we associate the graded vector space $\varinjlim_\alpha V_\alpha^\vee$. This construction is well defined: If $(\rho_\beta : V \rightarrow V_\beta)$ is an equivalent system for V, then for each α there are some β and a morphism $\varphi_{\beta\alpha}$ such that $\varphi_{\beta\alpha} \circ \rho_\beta = \rho_\alpha$,

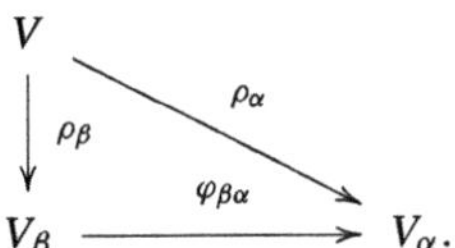

Since ρ_α is surjective, the same is true for $\varphi_{\beta\alpha}$. Therefore $V_\alpha^\vee$ injects into $V_\beta^\vee$, and $\varinjlim_\alpha V_\alpha^\vee$ is a subspace of $\varinjlim_\beta V_\beta^\vee$. Since we have also an inverse inclusion, the two spaces are equal.

Now, any linear map of graded vector spaces $\varphi : V = \cup V_\alpha \rightarrow W = \cup W_\beta$ dualizes to a morphism $W' = \varprojlim_\beta W_\beta^\vee \rightarrow V' = \varprojlim_\alpha V_\alpha^\vee$. Moreover, if $f : W' \rightarrow V'$ is a morphism of enriched graded vector spaces, then for any α there are some β and a commutative diagram

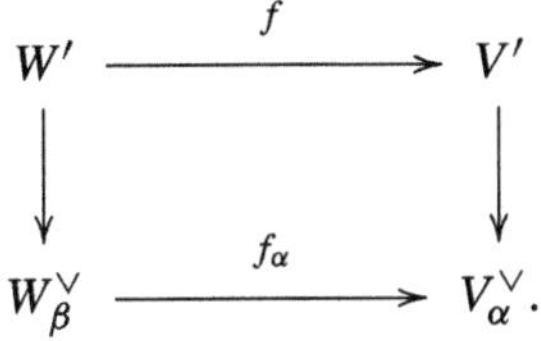

Dualizing the f_α gives a well-defined map $V \rightarrow W$. $\qquad\qquad\square$

Proposition 2.2

(i) *If $\varphi : S \to T$ is a morphism between enriched vector spaces, then $\mathrm{Im}\,\varphi$ and $\mathrm{Ker}\,\varphi$ are enriched vector subspaces.*

(ii) *Suppose (N, ∂) is an enriched $\mathbb{Z}$-graded chain complex, equipped with a decreasing filtration $N(q)$ for which $\cap_q N(q) = 0$. The corresponding spectral sequence is then convergent.*

Proof

(i) This is immediate from Proposition 2.1 and standard linear algebra.

(ii) Let $M = N^\vee$ and $M(q) = (N/N(q))^\vee$. The $M(q)$ provide an increasing filtration with $M = \cup_q M(q)$ so that the associated spectral sequence is convergent. The spectral sequence for N is the dual spectral sequence.

$\square$

Lemma 2.1 *Let $(V(r), f_{rs} : V(r) \to V(s))$ be an inverse system of morphisms of enriched vector spaces. Then $V = \varprojlim_r V(r)$ is an enriched vector space and is the limit of the system in the category of enriched vector spaces.*

Proof Denote by $(U_r, g_{rs} : U_s \to U_r)$ the dual of the diagram, and let $U = \varinjlim_r U_r$. Then $V = \varprojlim_r V_r = U^\vee$ is an enriched vector space, and a straightforward computation shows that this is the limit in the category of enriched vector spaces. $\square$

Lemma 2.2 *Suppose $S \subset T$, and $(T, \{\rho_\alpha\})$ is a pre-enriched vector space. Then completions preserve short exact sequences:*

$$\overline{T/S} = \varprojlim_\alpha \rho_\alpha T / \rho_\alpha S = \overline{(T/S)}.$$

In particular, if T is complete, then it follows that S is a subspace of $\overline{S}$, and the inclusion $S \to \overline{S}$ is the completion map. In this case we say $\overline{S}$ is the *closure* of S.

Proof Since $\rho_\alpha S \subset \rho_\alpha T$,

$$0 \to \rho_\alpha S \to \rho_\alpha T \to \rho_\alpha T / \rho_\alpha S \to 0$$

is a short exact sequence of finite dimensional spaces. Thus the lemma follows from Proposition 1.1. $\square$

Lemma 2.3 *Suppose $(T, \{\rho_\alpha\})$ is an enriched vector space. Then:*

(i) A finite sum of closed subspaces of T is closed.
(ii) An arbitrary intersection of closed subspaces of T is closed.
(iii) For any $S \subset T$, $\overline{S} = \cap_\alpha (S + \ker \rho_\alpha)$.
(iv) If $S \subset T$ is a graded space of finite type, then S is closed.
(v) Each closed subspace $S \subset T$ admits a closed direct summand.

Proof

(i) The same proof that for Lemma 1.2 (ii).
(ii) Suppose $x \in \cap_\sigma S_\sigma$, where each S_σ is a closed subspace of T. Then

$$\rho_\alpha x \in \rho_\alpha (\cap S_\sigma) \subset \cap_\sigma \rho_\alpha (S_\sigma).$$

Therefore $\rho_\alpha x \in \rho_\alpha (S_\sigma)$ for each σ. Thus $x = (\rho_\alpha x) \in \varprojlim_\alpha \rho_\alpha (S_\sigma) = \overline{S_\sigma} = S_\sigma$.
Thus $x \in \cap_\sigma S_\sigma$ and $\cap_\sigma S_\sigma = \overline{\cap_\sigma S_\sigma}$.
(iii) If $x \in \overline{S}$, then $\rho_\alpha x \in \rho_\alpha (S)$. Thus for some $x_\alpha \in S$,

$$\rho_\alpha (x - x_\alpha) = 0$$

and so $x - x_\alpha \in \ker \rho_\alpha$. Therefore $x \in S + \ker \rho_\alpha$ for all α.
   In the reverse direction, suppose $x \in \cap_\alpha (S + \ker \rho_\alpha)$. Then

$$\rho_\alpha x \in \rho_\alpha S.$$

Thus the coherent family $x = (\rho_\alpha x) \in \overline{S}$.
(iv) We suppose first that $S = \{x\}$. Let $y \in \overline{S}$. Then by (iii) $y - x \in \ker \rho_\alpha$ for all α. Therefore $y = x$ and $\overline{S} = S$. It follows from (i) that each S_k is closed and therefore that S is closed.
(v) Let $\rho_\alpha : T \to T_\alpha$ be the projections associated with T. We define R to be the inverse limit $\varprojlim_\alpha \overline{\rho_\alpha}$, where $\overline{\rho_\alpha}$ is the composition $T \to T_\alpha / \rho_\alpha (S)$. Then R is a closed subset, and the short exact sequences $0 \to \rho_\alpha (S) \to T_\alpha \to T_\alpha / \rho_\alpha (S) \to 0$ induce together the short exact sequence $0 \to S \to T \to R \to 0$.

$\qquad\qquad\qquad\qquad\qquad\qquad\qquad\qquad\qquad\qquad\qquad\qquad\qquad\qquad$ $\square$

Remark Suppose $(x_k)_{k \geq 1}$ is a sequence of elements in a pre-enriched vector space $(T, \{\rho_\alpha\})$. If for each α there is some $r(\alpha)$ such that $x_k \in \ker \rho_\alpha$, for $k > r(\alpha)$, then the sums $\sum_{k=1}^m x_k$ fit together to give a well-defined element, $\sum_{k=1}^\infty x_k$ in T.

2.2   Weight Decompositions

A *weight decomposition* in a graded Lie algebra is a decomposition

$$L = \oplus_{k \geq 0} L(k)$$

in which $[L(k), L(\ell)] \subset L(k + \ell)$. A *weighted subspace* of L is a subspace $S \subset L$ such that $S = \oplus_k S \cap L(k)$. We denote $S(k) = S \cap L(k)$.

A *weighted enriched Lie algebra* is a weighted Lie algebra with a defining set of ideals of the form $I_\alpha = \oplus_k I_\alpha(k)$.

Henceforth in this example, $L = \oplus_{k \geq 1} L(k)$ denotes a fixed weighted enriched Lie algebra with a defining set of weighted ideals I_α.

There follows the Proposition

Proposition 2.3 *With the hypotheses and notation above:*

(i) Each $L(k)$ is closed, and $\overline{L} = \prod_k L(k)$.

(ii) If $S \subset L$ is a weighted subspace, then $\overline{S} = \prod_k \overline{S(k)}$. In particular, $\prod_k S(k)$ is closed if and only if each $S(k)$ is closed, in which case any subspace of the form $\prod_i S(k_i)$ is closed.

(iii) If each $L(k)$ is finite dimensional, then $\overline{L} = \varprojlim_r L / \prod_{k > r} L(k)$.

Proof

(i) Denote as usual by $\rho_\alpha : L \to L_\alpha = L/I_\alpha$ the projection on the finite dimensional Lie algebra L_α. Then

$$L_\alpha = \oplus_k \rho_\alpha(L(k)) = \prod_k \rho_\alpha(L(k)),$$

and

$$L = \varprojlim_\alpha L_\alpha = \prod_k \varprojlim_\alpha \rho_\alpha(L(k)) = \prod_k \varprojlim_\alpha L(k)/I_\alpha(k) = \prod_k \overline{L(k)}.$$

(ii) This follows in the same way as (i). Then (iii) is a direct consequence.

$\square$

Example A free graded Lie algebra $\mathbb{L}(V)$ admits the weight decomposition given by $\mathbb{L}(V) = \oplus_k \mathbb{L}^k(V)$, where $\mathbb{L}^k(V)$ denotes the subvector space generated by the Lie brackets of length k in V. A more general example is given by the quotient $\mathbb{L}(V)/I$, where I is generated by quadratic relations in V.

Lower Central Series **3**

Throughout this section L denotes an enriched Lie algebra, with quotient maps $\rho_\alpha : L \to L_\alpha$. The lower central series $L^{(n)}$ of L is the analog for enriched Lie algebras of the usual lower central series L^n. The closed ideal $L^{(n)}$ is the closure of the ideal L^n.

A key invariant for any graded Lie algebra E is its *lower central series*

$$E = E^1 \supset \cdots \supset E^k \supset \cdots$$

of ideals in which E^k is the linear span of iterated commutators of length k of elements in E. The *classical completion* of E is the inverse limit

$$\widehat{E} = \varprojlim_n E/E^n,$$

and E is *pronilpotent* if $E \xrightarrow{\cong} \widehat{E}$.

The analog for the enriched Lie algebra L is the sequence of ideals

$$L = L^{(1)} \supset \cdots \supset L^{(k)} \supset \cdots$$

defined by $L^{(k)} = \overline{L^k}$: This is the *lower central series for an enriched Lie algebra*. It is immediate that $L \rightsquigarrow L^{(k)}$ is a functor. Moreover, since $L^k \subset L^{(k)}$, *each $L/L^{(k)}$ is a nilpotent Lie algebra:* $\left(L/L^{(k)}\right)^k = 0$.

Lemma 3.1 *Let L be an enriched Lie algebra. Then*

$$L = \varprojlim_n L/L^{(n)}.$$

Y. Félix, S. Halperin, *Lie Models for Spaces*, Frontiers in Mathematics,
https://doi.org/10.1007/978-3-032-15357-9_3

More precisely:

(i) $L^{(k)} = \varprojlim_{\ell} L^{(k)}/L^{(k+\ell)}$.
(ii) $[L^{(k)}, L^{(\ell)}] \subset L^{(k+\ell)}$.
(iii) L *is a retract of* $\widehat{L} = \varprojlim_{n} L/L^{n}$.

Proof

(i) Since each L_α is nilpotent, for some $\ell = \ell(\alpha)$, $L_\alpha^{k+\ell} = 0$. Thus by Lemma 2.2,

$$\varprojlim_{\ell} \overline{L^k}/\overline{L^{k+\ell}} = \varprojlim_{\ell,\alpha} L_\alpha^k/L_\alpha^{k+\ell} = \varprojlim_{\alpha} L_\alpha^k = \overline{L^k}.$$

(ii) From Lemma 2.3 we obtain $\rho_\alpha(L^{(k)}) = \rho_\alpha(\varprojlim_{\beta \geq \alpha} \rho_\beta(L^k)) = \rho_\alpha(L^k) = L_\alpha^k$. It follows that

$$\rho_\alpha[L^{(k)}, L^{(\ell)}] \subset [L_\alpha^k, L_\alpha^\ell] \subset L_\alpha^{k+\ell},$$

and so $[L^{(k)}, L^{(\ell)}] \subset \overline{[L^{(k)}, L^{(\ell)}]} \subset L^{(k+\ell)}$.

(iii) For each n and α, the surjection $L \to L_\alpha/L_\alpha^n$ (induced from ρ_α) factors as

$$L \to L/L^n \to L_\alpha/L_\alpha^n.$$

Thus we obtain

$$L \to \varprojlim_{n} L/L^n \to \varprojlim_{n,\alpha} L_\alpha/L_\alpha^n = \varprojlim_{\alpha} L_\alpha = L,$$

which decomposes id_L as $L \xrightarrow{\varphi} \widehat{L} \xrightarrow{\psi} L$. $\square$

Corollary *A morphism $E \to F$ of enriched Lie algebras is an isomorphism if and only if it induces isomorphisms $E/E^{(n)} \xrightarrow{\cong} F/F^{(n)}$, $n \geq 2$.*

Remark Lemma 3.1 identifies the sequence $L \supset \cdots \supset L^{(k)} \supset \ldots$ as an N-suite as defined by Lazard in [57].

Example As follows from Lemma 3.1, the direct sum

$$g(L) := \oplus_{n \geq 0} L^{(n)}/L^{(n+1)}$$

is an enriched weighted Lie algebra, with Lie bracket

$$\left(L^{(n)}/L^{(n+1)}\right) \times \left(L^{(k)}/L^{(k+1)}\right) \to L^{(n+k)}/L^{(n+k+1)}$$

induced from the Lie bracket in L.

Lemma 3.2 *Suppose $E \subset L$ is a sub-Lie algebra of the enriched Lie algebra L.*

  (i) If $E + L^{(2)} = L$, then $\overline{E} = L$.
  (ii) $\overline{E^n} = \overline{E}^{(n)}$, $n \geq 1$.
 (iii) If E is abelian (respectively, solvable), then $\overline{E}$ is abelian (respectively, solvable).

Proof

  (i) By hypothesis $\rho_\alpha L = \rho_\alpha E + \rho_\alpha L^{(2)} = \rho_\alpha E + [\rho_\alpha L, \rho_\alpha L]$. Since $\rho_\alpha L$ is nilpotent, we get by induction $\rho_\alpha L \subset \rho_\alpha E$ and $\overline{E} = L$.
 (ii) Since (Lemma 2.2) $E \subset \overline{E}$, $\rho_\alpha(\overline{E}) = \rho_\alpha(E)$. Therefore

$$\begin{aligned}
\overline{E^n} &= \varprojlim_\alpha \rho_\alpha(E^n) = \varprojlim_\alpha (\rho_\alpha(E))^n \\
&= \varprojlim_\alpha (\rho_\alpha(\overline{E}))^n = \varprojlim_\alpha \rho_\alpha(\overline{E}^n) = \overline{E}^{(n)}.
\end{aligned}$$

 (iii) This follows from (ii). $\qquad\qquad\square$

Proposition 3.1 *If for each integer p, $\dim(L/L^{(2)})_p < \infty$, then:*

  (i) For any p and any integer k, there is an α such that the projection $(L/L^{(k)})_p \to (L_\alpha/L_\alpha^k)_p$ is an isomorphism.
 (ii) $L^k = L^{(k)}$, $k \geq 1$, and so L is pronilpotent.

Proof

  (i) By Lemma 2.2 there is an α_0, such that for each $\beta \geq \alpha_0$ the projection $p_\beta : L/L^{(2)} \to L_\beta/L_\beta^2$ is an isomorphism in degree p. This shows that for each α the dimension of L_α/L_α^2 in degree p is bounded by the integer $N = \dim(L_{\alpha_0}/L_{\alpha_0}^2)_p$. It follows that for any $r \geq 1$ and any α the dimension of $L_\alpha^r/L_\alpha^{r+1}$ in degree p is bounded by N^r. Thus it follows from Lemma 2.2 that $L^{(r)}/L^{(r+1)} = \varprojlim_\alpha L_\alpha^r/L_\alpha^{r+1}$ is finite dimensional in degree p. Therefore for any integer k there is an index α_k such that the projection

$$(L/L^{(k)})_p \to (L_{\alpha_k}/L_{\alpha_k}^k)_p$$

 is an isomorphism.

(ii) It follows from Lemma 2.2 that, given p, for some α the Lie bracket in L

$$L/L^{(2)} \otimes L^{(r)}/L^{(r+1)} \to L^{(r+1)}/L^{(r+2)}, \qquad r \geq 1, \tag{3.1}$$

in degree p may be identified with the corresponding surjection

$$L_\alpha/L_\alpha^2 \otimes L_\alpha^r/L_\alpha^{r+1} \to L_\alpha^{r+1}/L_\alpha^{r+2}.$$

This implies that the Lie bracket (3.1) is surjective in each degree p.

Now let $x_1, \ldots, x_m \in L$ represent a basis of $(L/L^{(2)})_{\leq p}$. Then for any $x \in L_p^{(n)}$ there are elements $y_i \in L^{(n-1)}$ such that

$$y - \sum_{i=1}^m [x_i, y_i] \in L^{(n+1)}.$$

This yields an inductive construction of elements $y_i(\ell) \in L^{(n+\ell)}$, $\ell \geq -1$, such that

$$x - \sum_{i=1}^m \sum_{\ell=-1}^r [x_i, y_i(\ell)] \in L^{(n+r+2)}.$$

Set $y_i = \sum_\ell y_i(\ell)$. Then by construction

$$x = \sum_{i=1}^m [x_i, y_i].$$

This shows that $L^{(k)} \subset [L, L^{(k-1)}]$. Induction on k gives

$$L^{(k)} \subset L^k,$$

and the reverse inclusion is obvious. $\qquad\qquad\qquad\qquad\qquad\qquad\qquad\qquad\qquad\square$

Corollary 3.1 *Let E be a pre-enriched Lie algebra with $\dim E/E^2 < \infty$. Then $\dim \overline{E}/\overline{E}^k < \infty$, $k \geq 1$, and the lower central series for $\overline{E}$ satisfies*

$$\overline{E}^{(k)} = \overline{E}^k = \overline{E^k}.$$

Proof There is an α_0 such that for each $\alpha \geq \alpha_0$, the projection $E_\alpha/E_\alpha^2 \to E_{\alpha_0}/E_{\alpha_0}^2$ is an isomorphism. Thus by Lemma 2.2 $\overline{E}/\overline{E}^{(2)}$ is isomorphic to $E_{\alpha_0}/E_{\alpha_0}^2$, and so is finite dimensional. Proposition 3.1 gives then an isomorphism $\overline{E}/\overline{E}^{(2)} = \overline{E}/\overline{E}^2 = E_{\alpha_0}/E_{\alpha_0}^2$. In particular, $E + \overline{E}^{(2)} = \overline{E}$. The result is then a direct consequence of Proposition 3.1(ii) and Lemma 3.2(ii).

$\square$

Corollary 3.2 *An enriched Lie algebra L is the direct limit of its closed sub-Lie algebras, E, satisfying dim $E/E^2 < \infty$.*

Proof Any Lie algebra is the direct limit of its finitely generated Lie algebras, F. Thus L is the direct limit of the completions, $\overline{F}$. By Proposition 3.1 and Corollary 3.1, each F satisfies dim $\overline{F}/\overline{F}^2 < \infty$.

$\square$

Proposition 3.1 has the following analog:

Proposition 3.2 *If $E \subset L$ is a sub-Lie algebra and E/E^2 is a graded vector space of finite type, then for each k:*

(i) $E/E^k \stackrel{\cong}{\to} \overline{E}/\overline{E^k}$.
(ii) $\overline{E}^k = \overline{E^k} = \overline{E}^{(k)}$.
(iii) $\widehat{E} = \overline{E} = \widehat{\overline{E}}$.

In particular, $\overline{E}$ is pronilpotent.

Proof

(i) Since E/E^2 has finite type, each E/E^k has finite type. Thus (i) follows from Lemma 2.2.
(ii) This follows exactly as in Corollary 3.1 to Proposition 3.1.
(iii) This follows immediately from (i) and (ii).

$\square$

Proposition 3.3 *Suppose L is an enriched Lie algebra such that L/L^2 has finite type. Then L has a unique (up to equivalence) structure as an enriched Lie algebra $(L, \mathfrak{I})$, namely that given by*

$$\mathfrak{I} = \{I_k\}_{k \geq 1} \quad \text{with } I_k = L^k + L_{\geq k}.$$

Proof Suppose a second family $\mathfrak{J} = \{J_\beta\}$ of ideals in L also makes L into an enriched Lie algebra. Since L/J_β is finite dimensional and nilpotent, it is also immediate that $J_\beta \supset I_k$ for some k.

In the reverse direction, apply Proposition 3.1(i) to conclude that for fixed k and n there is some $\beta = \beta(k, n)$ such that

$$\rho_\beta : L_n/(L^k)_n \xrightarrow{\;\cong\;} (L_\beta)_n/(L_\beta^k)_n.$$

It follows that

$$(L^k)_n = \rho_\beta^{-1}(L_\beta^k)_n \supset (J_\beta)_n.$$

Choose $\gamma \in \mathcal{J}$ so that

$$J_\gamma \subset \cap_{n=1}^{k} J_{\beta(k,n)}.$$

Then

$$(L^k)_{\leq n} \supset (J_\gamma)_{\leq n},$$

and so

$$I_k \supset J_\gamma.$$

$$\square$$

Corollary *Let L be an enriched Lie algebra and $E \subset L$ be a subalgebra such that E/E^2 has finite type. Then E is closed if and only if E is pronilpotent.*

Proof By Proposition 3.3, the structure systems $\rho_\alpha E$ and E/E^n are equivalent. This implies the result. $\square$

Example Suppose a sub-Lie algebra, $E \subset L$, satisfies

$$\cap_k E^k = 0, \quad \dim E/E^2 < \infty, \quad \text{and} \dim E = \infty.$$

Let $z \in \overline{E}$ satisfy $E \oplus \mathbb{Q}z \subset \overline{E}$, and let F be the sub-Lie algebra generated by E and $\mathbb{Q}z$. Then $E \subset F \subset \overline{E}$, and it follows that $\overline{E} = \overline{F}$. However the dimension of F/F^2 depends on the choice of z.

For instance, suppose E is the completion of the free Lie algebra $\mathbb{L}(x, y)$ generated by x, y, with the enriched structure described in Sect. 1.1. Let F be the sub-Lie algebra generated by x, y, and $z = e^{ad\,x}(y)$. Since t and e^t are algebraically independent in $\mathbb{Q}[[t]]$, the classes of x, y, and z are linearly independent in F/F^2, and F/F^3 has dimension 3.

On the other hand, with the same E, if $z = \sum_{n\geq 0} \mathrm{ad}_x^n(y)$, then $z - y = [x, z]$ and $F/F^2 = E/E^2 = \mathbb{Q}x \oplus \mathbb{Q}y$.

3.1 The Fundamental Group G_L of an Enriched Lie Algebra L

Associated with an enriched Lie algebra, L, is a group G_L defined as follows: For each α, the diagonal map $L_\alpha \to L_\alpha \times L_\alpha$ extends to a morphism

$$\Delta_\alpha : \widehat{U}L_\alpha \to \widehat{U}L_\alpha \,\widehat{\otimes}\, \widehat{U}L_\alpha := \widehat{U}(L_\alpha \times L_\alpha).$$

An element $x \in \widehat{U}L_\alpha$ is called *group-like* if $\Delta_\alpha x = x \otimes x$, and the group-like elements form a group G_α with multiplication given by the product in the algebra $\widehat{U}L_\alpha$.

Definition The *fundamental group*, G_L, of an enriched Lie algebra, L, is the group

$$G_L = \varprojlim_\alpha G_\alpha.$$

Note that, since each L_α is finite dimensional and nilpotent, the classical exponential map $L_\alpha \to \widehat{U}L_\alpha$ is well defined, and it is straightforward to verify that it is a bijection onto G_α. The inverse is then given by the standard power series for log, which is trivially also convergent. Passing to inverse limits this yields bijections which we denote by

$$L \underset{\log}{\overset{\exp}{\rightleftarrows}} G_L.$$

On the other hand, the classical lower central series for a group G is the sequence (G^n) of subgroups generated by iterated commutators $[a_1, [a_2, \ldots [a_{n-1}, a_n] \ldots]$ of length n. In analogy with the series $L^{(n)}$, we define

$$G_L^{(n)} = \varprojlim_\alpha G_\alpha^n.$$

In particular it follows that $\exp$ and $\log$ restrict to bijections

$$L^{(n)} \underset{\log}{\overset{\exp}{\rightleftarrows}} G_L^{(n)}$$

since these are established for each L_α, G_α in [36, Theorem 2.2]. Then in view of Proposition 1.1 these factor to yield bijections

$$L^{(n)}/L^{(k)} \overset{\cong}{\longrightarrow} G_L^{(n)}/G_L^{(k)}$$

since the corresponding bijections for each L_α are established in [36, Proposition 2.5]. In particular [36, Corollary 2.4], for each n, the bijection

$$L^{(n)}/L^{(n+1)} \xrightarrow{\;\cong\;} G_L^{(n)}/G_L^{(n+1)}$$

is an isomorphism of rational vector spaces.

The Quadratic Sullivan Model of an Enriched Lie Algebra **4**

By a generalization of the standard cochain construction, we associate a quadratic Sullivan algebra $(\wedge V, d)$ with any enriched Lie algebra L with the property that the suspension of L is isomorphic to $\mathrm{Hom}(V, \mathbb{Q})$. The relations between L and its quadratic Sullivan model will be very useful in the next chapters.

A *quadratic Sullivan algebra* is a Sullivan algebra $(\wedge V, d)$ in which $d : V \to \wedge^2 V$. This endows $H(\wedge V)$ with the decomposition

$$H(\wedge V) = \oplus_k H^{[k]}(\wedge V)$$

in which $H^{[k]}(\wedge V)$ is the subspace represented by $\wedge^k V \cap \ker d$. Moreover, $V = \varinjlim_n V_n$, where

$$V_0 = V \cap \ker d \quad \text{and} \quad V_{n+1} = V \cap d^{-1}(\wedge^2 V_n).$$

Then a *morphism* $\varphi : \wedge V \to \wedge W$ of quadratic Sullivan algebras is a cdga morphism which restricts to a linear map $V \to W$. Thus it restricts to linear maps

$$\varphi_n : V_n \to W_n.$$

The filtration $(V_n)_{n \geq 0}$ will be called the *standard filtration*.

It is straightforward to verify that any quadratic Sullivan algebra $(\wedge V, d)$ decomposes as

$$(\wedge V, d) = \varinjlim_\alpha (\wedge V_\alpha, d_\alpha)$$

in which the V_α are the finite dimensional subspaces of V for which $d(V_\alpha) \subset \wedge^2 V_\alpha$.

Our first objective here is to construct a contravariant isomorphism from the category of enriched Lie algebras to the category of quadratic Sullivan algebras. For this, recall that the classical functor $E \rightsquigarrow C_*(E)$ from graded Lie algebras to cocommutative chain coalgebras is given by $C_*(E) = \wedge sE$, with differential determined by the condition

$$\partial(sx \wedge sy) = (-1)^{\deg x+1} s[x, y].$$

The dual $C^*(E) = \mathrm{Hom}(C_*(E), \mathbb{Q})$ is, with the differential forms on a manifold, one of the earliest examples of a cdga. In particular, if $\dim E < \infty$, then $C^*(E) = \wedge(sE)^\vee$.

Now set $V_\alpha = (sL_\alpha)^\vee$. Then $C_*(L_\alpha) = (\wedge V_\alpha, d_\alpha)$, and, since L_α is finite dimensional and nilpotent, $(\wedge V_\alpha, d_\alpha)$ is a quadratic Sullivan algebra in which

$$< d_\alpha v, sx, sy > = (-1)^{\deg y+1} < v, s[x, y] >, v \in V_\alpha, x, y \in L_\alpha. \qquad (4.1)$$

Definition $(\wedge V, d) = \varinjlim_\alpha (\wedge V_\alpha, d_\alpha)$ is the *quadratic Sullivan model* of $(L, \mathfrak{J})$.

In the reverse direction, suppose $(\wedge V, d)$ is a quadratic Sullivan algebra, and let $\{V_\alpha \subset V\}$ be the family of finite dimensional subspaces for which $d : V_\alpha \to \wedge^2 V_\alpha$. Then (4.1) defines an inverse system of finite dimensional nilpotent Lie algebras L_α, and $L = \varprojlim_\alpha L_\alpha$ is an enriched Lie algebra with surjections $\rho_\alpha : L \to L_\alpha$ (Proposition 2.1).

Definition $(L, \{\rho_\alpha\})$ is the *homotopy Lie algebra* of $(\wedge V, d)$.

It is immediate from the constructions above that they are (up to isomorphism) inverse to each other. In particular, if $(L, \mathfrak{J})$ and $(\wedge V, d)$ correspond to each other, then

$$\left. \begin{array}{c} sL = \varprojlim_\alpha sL_\alpha = \varprojlim_\alpha V_\alpha^\vee = V^\vee, \quad \text{and} \\[2mm] < dv, sx, sy > = (-1)^{1+\deg y} < v, s[x, y] >, \quad v \in V, x, y \in L. \end{array} \right\} \qquad (4.2)$$

Definition The *generalized cochain algebra* $\widehat{C}(L)$ on an enriched Lie algebra $L = \varprojlim_\alpha L_\alpha$ is the direct limit

$$\widehat{C}(L) = \varinjlim_\alpha C^*(L_\alpha).$$

Finally, we complete our objective with the following proposition:

Proposition 4.1 *The correspondence $(L, \mathfrak{J}) \rightsquigarrow \widehat{C}(L) = \wedge V$ is a contravariant isomorphism from the category of enriched Lie algebras to the category of quadratic Sullivan algebras.*

Proof It follows from Proposition 2.1 that morphisms of enriched Lie algebras $\varphi : E \to L$ are exactly the duals of linear maps $\psi : V_E \leftarrow V_L$ between the duals. It follows from (4.1) that φ is a morphism of Lie algebras if and only if ψ extends to a morphism $\wedge V_E \leftarrow \wedge V_L$ of quadratic Sullivan algebras. $\qquad\square$

Remarks

1. It follows from (4.2) that the enriched structure of L is defined by the isomorphism $sL = V^\vee$.
2. The natural morphism $\wedge V = \varinjlim C^*(L_\alpha) \to C^*(L)$ will not, in general, be a quasi-isomorphism. For instance, if $V = V^1$ has zero differential and countably infinite dimension, then $\wedge V$ is countable, but L and $C^*(L)$ are uncountable, and since L is abelian, $C^*(L) = H(C^*(L))$.

The correspondence of Proposition 4.1 is reflected in the next lemma.

Lemma 4.1

(i) *The identification* $sL \xrightarrow{\cong} V^\vee$ *induces natural isomorphisms*

$$s\left(L/L^{(n+2)}\right) \xrightarrow{\cong} V_n^\vee, n \geq 0,$$

and therefore identifies $L^{(n+2)} = \{x \in L \mid\, < V_n, sx >\, = 0\}$.

(ii) *If* $T \oplus L^{(2)} = L$, *then* L *is the closure of the sub-Lie algebra,* E, *generated by* T.

Proof

(i) This is straightforward when L and V are replaced by L_α and V_α. In the general case by Lemma 2.2 we have the following sequence of isomorphisms:

$$s\left(L/L^{(n+2)}\right) = s\left(\overline{L}/\overline{L^{n+2}}\right) = s\left(\varprojlim_\alpha \rho_\alpha(L)/\rho_\alpha(L^{n+2})\right) = \varprojlim_\alpha s\left(L_\alpha/L_\alpha^{n+2}\right)$$

$$= \varprojlim_\alpha (V_\alpha)_n{}^\vee = \left(\varinjlim_\alpha (V_\alpha)_n\right)^\vee = V_n^\vee.$$

(ii) The inclusion $sE \to sL$ is the dual of a surjection $W \xleftarrow{\varphi} V$ where $\wedge W$ is the quadratic model of E. By construction $E/E^{(2)} \xrightarrow{\cong} L/L^{(2)}$, and so $W_0 \xleftarrow{\cong} V_0$. It remains to prove that φ is injective. Suppose that $W_n \xleftarrow[\cong]{\varphi} V_n$, and let $v \in V_{n+1}$ with $\varphi v = 0$. Then $\varphi dv = 0$, so that $dv = 0$ and $v \in V_0$, which implies that $v = 0$. $\qquad\square$

Remark A morphism $\psi : E \to L$ of enriched Lie algebras induces morphisms $\psi(n) :$ $E/E^{(n)} \to L/L^{(n)}$, $n \geq 1$. If $\varphi : \wedge V_L \to \wedge V_E$ is the corresponding morphism of quadratic models, then $\psi(n+2)$ is dual to the linear map $\varphi_n : (V_L)_n \to (V_E)_n$.

Lemma 4.2 *The closure of any subspace $S \subset L$ is the image of the inclusion $(V/K)^\vee \to V^\vee = sL$, where*

$$K = \{v \in V \mid \; < v, sS >= 0\}.$$

In particular, any closed subspace of L has a closed direct summand.

Proof For $x \in L$ and $v \in V_\alpha$, we have $< v, sx >=< v, s\rho_\alpha x >$. Thus

$$V_\alpha \cap K = \{v \in V_\alpha \mid \; < v, s\rho_\alpha S >= 0\}.$$

Set $P_\alpha = V_\alpha / V_\alpha \cap K$. Then, because V_α and L_α are finite dimensional, this gives

$$\rho_\alpha S = \{x \in L_\alpha \mid \; < V_\alpha \cap K, sx >= 0\} = (sP_\alpha)^\vee.$$

Thus denoting $V/K = \varinjlim_\alpha P_\alpha = P$, we obtain

$$\overline{S} = \varprojlim_\alpha \rho_\alpha S = \varprojlim_\alpha (sP_\alpha)^\vee = (\varinjlim_\alpha sP_\alpha)^\vee = (sP)^\vee.$$

Finally the inclusion $S \to L$ is the dual of a surjection $V \to V/K$. Dualizing the inclusion $K \to V$ provides a surjection $L \to (sK)^\vee$ onto a closed direct summand of S. $\qquad\qquad\square$

Example (Weighted Lie Algebras) Let $E = \oplus_k E(k)$ be a weighted enriched Lie algebra with defining ideals $I_\alpha = \oplus_k I_\alpha(k)$. The weighting then induces another gradation $V = \oplus_k V(k)$ in the generating space of the corresponding quadratic model:

$$V(k) = \varinjlim_\alpha s\, (E(k)/I_\alpha(k))^\vee .$$

It is immediate that the differential preserves the induced gradation in $\wedge V$.

4.1　　Closed Subalgebras and Ideals

Suppose $E \subset L$ is a closed subalgebra of an enriched Lie algebra. Then the inclusion corresponds via duality to a surjection $\wedge V \xrightarrow{\;\rho\;} \wedge Z$ of the respective quadratic models, and by Lemma 4.2,

$$E = \{x \in L \mid\ < \ker \rho_{|V}, sx >= 0\}.$$

Moreover, $d : \ker \rho_{|V} \to \ker \rho_{|V} \wedge V$ [36, Lemma 10.4]. Recall also that every such surjection induces an inclusion of a closed subalgebra of L.

Now suppose $I \subset L$ is a closed ideal. Let $\rho : \wedge V \to \wedge Z$ be the corresponding surjection, and denote $W = \ker \rho_{|V}$. In this case the condition that I is an ideal is equivalent to the condition $d : W \to \wedge^2 W$ [36, Lemma 10.4]. Thus $\wedge W$ is a subquadratic Sullivan algebra, and I decomposes $\wedge V$ as the Sullivan extension

$$\wedge W \xrightarrow{\ \lambda\ } \wedge W \otimes \wedge Z = \wedge V \xrightarrow{\ \rho\ } \wedge V \otimes_{\wedge W} \mathbb{Q} = \wedge Z,$$

which dualizes to

$$L/I \leftarrow L \leftarrow I.$$

In particular this identifies L/I as the homotopy Lie algebra of $\wedge W$.

Proposition 4.2 *Let $\varphi : E \to L$ be a morphism of enriched Lie algebras.*

(i) Both $\mathrm{Ker}\,\varphi$ and $\mathrm{Im}\,\varphi$ are closed subspaces, respectively, in E and L.
(ii) If φ induces a surjection $E/E^{(2)} \to L/L^{(2)}$, then φ is surjective.

Proof

(i) is a special case of Proposition 2.1 (iii).
(ii) This follows from Lemma 3.2(i)   □

Corollary *Let $\varphi : E \to L$ be a surjective morphism between enriched abelian Lie algebras. Then φ admits a section, σ.*

Proof By Proposition 4.2, $\mathrm{Ker}\,\varphi$ is a closed subspace. Now by Lemma 2.3 (v), $\mathrm{Ker}\,\varphi \subset L$ admits a closed direct summand S. It follows that $\varphi_{|S} : S \to \mathrm{Im}\,\varphi$ is an isomorphism, and we define $\sigma = (\varphi_{|S})^{-1}$.   □

Proposition 4.3 *Suppose $\{L(\sigma), \varphi_{\sigma,\tau} : L(\tau) \to L(\sigma)\}$ is an inverse system of morphisms of enriched Lie algebras, and suppose that either the morphisms are surjective or the system is finite. Then*

$$L := \varprojlim_{\sigma} L(\sigma)$$

is naturally an enriched Lie algebra.

Proof Suppose first that the morphisms are surjective. Under the correspondence of Proposition 4.1, the inverse system is the dual of a directed system $\{\wedge V(\sigma)\}$ of injections between the quadratic Sullivan models of the $L(\sigma)$. But then $\wedge V := \varinjlim_\sigma \wedge V(\sigma)$ is a quadratic Sullivan algebra, and L is the homotopy Lie algebra of $\wedge V$. It is immediate that L is the inverse limit in the category of enriched Lie algebras.

When the system is finite, we replace the system $L(\sigma)$ by the subsystem formed by the $G(\sigma)$ where $G(\tau)$ is the intersection of the images of morphisms $L(\sigma) \to L(\tau)$, coming from all $L(\sigma)$. By Proposition 4.2 the images are closed subalgebras, and by Lemma 2.2 their intersection is also a closed sub-Lie algebra. Now by construction the induced maps between the $G(\sigma)$ are surjective. Since $\varprojlim_\sigma L(\sigma) = \varprojlim_\sigma G(\sigma)$, the inverse limit is an enriched Lie algebra.

$\square$

Remark It is consistent with the ZFC axioms that the same graded Lie algebra can support two inequivalent enriched structures.

In fact, let $L = L_0$ be a vector space, and suppose (consistent with the ZFC axioms, [76]) that there are isomorphisms

$$V^\vee \cong sL \cong W^\vee$$

in which card $V \neq$ card W. Let $\wedge V$ and $\wedge W$ be the quadratic Sullivan algebras with zero differential. Then L regarded as an abelian Lie algebra is the homotopy Lie algebra of both $\wedge V$ and $\wedge W$, but since $\wedge V \not\cong \wedge W$ the corresponding enriched structures in L are not equivalent.

Remark that it is also consistent with ZFC to admit the Generalized Continuum Hypothesis (GCH) with the consequence that an isomorphism $V^\vee \cong W^\vee$ implies an isomorphism $V \cong W$ [55].

Representations of Enriched Lie Algebras

5

In this section L denotes an enriched Lie algebra and $\wedge V$ its quadratic Sullivan model. We proceed to a detailed study of the properties of L-modules and their corresponding $\wedge V$-modules. A special interest will be given to the profree L-models $V \widehat{\otimes} \overline{UL}$ associated with an enriched vector space V.

First recall that the left and right adjoint representations of L in L are given by

$$\mathrm{ad}_y(x) = [y, x] \quad \text{and} \quad \mathrm{ad}_y^{\mathrm{r}}(x) = (-1)^{\deg y}[x, y]. \tag{5.1}$$

As well, if $L \otimes M \to M$ is any representation of L (M is a *left L-module*), then the dual, $M^\vee$, of M inherits the right L-module structure given by

$$< m, a \cdot y >= (-1)^{\deg y} < y \cdot m, a >, \qquad m \in M, a \in M^\vee, y \in L. \tag{5.2}$$

Definition

(i) An *elementary L-module* is a finite dimensional left (respectively, right) nilpotent L-module, Q, for which some $I_\alpha \cdot Q = 0$ (respectively, $Q \cdot I_\alpha = 0$).

(ii) An *enriched L-module* is an L-module, N, together with a decomposition,

$$N = \varprojlim_\tau N_\tau,$$

of N as an inverse limit of elementary L-modules.

(iii) A *locally elementary L-module* is an L-module, M, together with a decomposition, $M = \varinjlim_\sigma M_\sigma$, of M as a direct limit of elementary L-modules.

© The Author(s), under exclusive license to Springer Nature Switzerland AG 2026

Y. Félix, S. Halperin, *Lie Models for Spaces*, Frontiers in Mathematics,

https://doi.org/10.1007/978-3-032-15357-9_5

(iv) A *morphism of enriched (respectively, locally elementary) L-modules* is a morphism of L-modules which is the inverse limit (respectively, direct limit) of morphisms of elementary L-modules.

Examples

(1) L with the adjoint representation and its universal enveloping algebra $\overline{UL}$ are natural examples of enriched L-modules.
(2) Suppose L is the free Lie algebra on one generator x in degree 0. The space M of finite sequences of rational numbers $(a_1, \ldots, a_n)$ equipped with the L-structure defined by $x \cdot (a_1, \ldots, a_n) = (a_2, \ldots a_n)$ is a locally elementary L-module. On the other hand the space N of infinite sequences $(a_1, a_2, \ldots)$ with the L-structure defined by $x \cdot (a_1, a_2, \ldots) = (0, a_1, a_2, \ldots)$ is an enriched L-module.

Proposition 5.1

(i) *The correspondence $M \rightsquigarrow M^\vee$ is a contravariant isomorphism from the category of locally elementary left L-modules to the category of enriched right L-modules.*
(ii) *Both the left and right adjoint representations of L in L convert L into an enriched L-module.*
(iii) *If M is either a locally elementary or an enriched L-module, then that representation of L extends uniquely to a representation of $\overline{UL}$.*
(iv) *Let $(\wedge V, d)$ be a quadratic Sullivan algebra. The isomorphism $sL = V^\vee$ identifies the right adjoint representation of L with the dual of the locally elementary representation in V given by $< y \cdot v, sx > = - < dv, sx, sy >, y, x \in L, v \in V,$*

$$\langle y \cdot v, sx \rangle = \langle v, s\, ad_y^r(x) \rangle.$$

Proof (i), (ii), and (iii) follow immediately from the definitions and the enriched structure $L = \varprojlim_\alpha L_\alpha$, in which L_α is finite dimensional and nilpotent. Finally, (iv) follows from this and (4.1):

$$\langle dv, sx, sy \rangle = (-1)^{\deg y + 1} \langle v, s[x, y] \rangle = -\langle v, s\, ad_y^r(x) \rangle.$$

$\square$

5.1   Standard Constructions for L-Modules

(1) If M and N are, respectively, a locally elementary and an enriched L-module, then $\mathrm{Hom}(M, N)$ is an enriched L-module.

(2) The sub- and quotient modules of a locally elementary (respectively, enriched module) are locally elementary (respectively, enriched). (The second assertion follows from Proposition 5.1.)

(3) The tensor product of locally elementary modules is locally elementary.

(4) The adjoint representation of L in any closed ideal I is an enriched representation.

Remarks

1. When $\dim L/[L, L] < \infty$, any finite dimensional nilpotent L-module is elementary, but this is not true in general. In fact, if $L = L_0$ is an infinite dimensional abelian Lie algebra with basis $\{x_i\}$, the subspaces $I(k)$ spanned by the x_i, $i \geq k$, define an enriched structure in L. Then, the finite dimensional nilpotent L-module $M = \mathbb{Q}a \oplus \mathbb{Q}b$ defined by

$$x_i \cdot a = b \quad \text{and} \quad x_i \cdot b = 0$$

   is not an elementary L-module.

2. Since for each elementary left L-module Q there is an ideal I_α with $I_\alpha \cdot Q = 0$, the module Q is naturally a $\widehat{UL_\alpha}$-module for some α. Thus the representations of L in enriched L-modules naturally extend to representations of the algebra $\overline{UL}$.

3. The surjections $\overline{UL} \to \widehat{UL_\alpha}/I_\alpha^n$ make $\overline{UL}$ into a right enriched L-module under right multiplication.

4. Let N be a right enriched L-module. The space of *decomposable elements* of N is by definition $\overline{N \cdot L} = \varprojlim_\alpha N_\alpha \cdot L$. For instance, for $N = L$ considered as a right enriched module over itself via the adjoint representation, we have $\overline{N \cdot L} = L^{(2)}$.

Augmentations

If $N = \varprojlim_\tau N_\tau$ is a right enriched L-module, the trivial L-modules, $\widetilde{N}_\tau := N_\tau \otimes_{\overline{UL}} \mathbb{Q} = N_\tau/N_\tau \cdot L$, define the trivial right enriched L-module, $\widetilde{N} = \varprojlim_\tau \widetilde{N}_\tau$ (i.e., $\widetilde{N}$ is an enriched vector space).

Definition The *augmentation of a right enriched L-module* is the induced morphism

$$\varepsilon_N : N \to \widetilde{N}$$

By construction, $\varepsilon_N : N \to \widetilde{N}$ depends naturally on N.

Let M and $\widetilde{M}$ be the locally elementary L-modules defined by $M^\vee = N$ and $\widetilde{M}^\vee = \widetilde{N}$. Then $\widetilde{M} = \{m \in M \mid L \cdot m = 0\}$.

5.2   Quadratic $(\wedge V, d)$-Modules and Holonomy Representations

Definition

(i) A *quadratic* $(\wedge V, d)$-module is a $(\wedge V, d)$-module of the form $\wedge V \otimes M$ in which
$d : 1 \otimes M \to V \otimes M$ and $M = \cup_{k \geq 0} M_k$, where

$$M_0 = M \cap \operatorname{Ker} d \quad \text{and} \quad M_{k+1} = M \cap d^{-1}(V \otimes M_k).$$

A *morphism* of quadratic $(\wedge V, d)$-modules is a morphism of the form $id \otimes \varphi : \wedge V \otimes M \to \wedge V \otimes M'$.

(ii) The *holonomy representation* for a quadratic $(\wedge V, d)$-module $\wedge V \otimes M$ is the left representation of L in M given by

$$x \cdot m = -\sum < v_i, sx > m_i,$$

where $d(1 \otimes m) = \sum v_i \otimes m_i$. In other words, the action $L \otimes M \to M$ is the composition

$$L \otimes M \xrightarrow{\;\; s \otimes d \;\;} sL \otimes V \otimes M \xrightarrow{\;\; <\,> \otimes \, id_M \;\;} \mathbb{Q} \otimes M = M.$$

Lemma 5.1

(i) The correspondence $(\wedge V \otimes M, d) \rightsquigarrow M$ is a natural isomorphism between the category of quadratic $\wedge V$-modules and locally elementary L-modules.

(ii) A morphism $(id \otimes \varphi) : \wedge V \otimes M \to \wedge V \otimes M'$ is a quasi-isomorphism if and only if $\varphi : M \to M'$ is an isomorphism.

Proof

(i) It is immediate that the holonomy representation makes M into a locally elementary left L-module. In the reverse direction, suppose $M = \varinjlim_\sigma M_\sigma$ is a locally elementary left L-module. The classical Cartan-Eilenberg-Serre construction then has the form

$$C^*(L_{\alpha(\sigma)}, M_\sigma) = \wedge (s L_{\alpha(\sigma)})^\vee \otimes M_\sigma = \wedge V_{\alpha(\sigma)} \otimes M_\sigma$$

in which $d : M_\sigma \to V_{\alpha(\sigma)} \otimes M_\sigma$. Passing to direct limits constructs the quadratic $\wedge V$-module $\wedge V \otimes M = \varinjlim_\sigma \wedge V_{\alpha(\sigma)} \otimes M_\sigma$.

(ii) The wedge gradation of $\wedge V$ makes a quadratic $\wedge V$-module into a chain complex, and, if $id \otimes \varphi$ is a quasi-isomorphism,

$$M \cap \ker d = H^{[0]}(\wedge V \otimes M) \xrightarrow{\cong} H^{[0]}(\wedge V \otimes M') = M' \cap \ker d.$$

Now suppose by induction that φ restricts to an isomorphism $M_k \xrightarrow{\cong} M_k'$. If $x \in M_{k+1}$ and $\varphi x = 0$, then $(id \otimes \varphi)dx = 0$. Since $dx \in V \otimes M_k$, it follows that $dx = 0$. Thus $x \in M \cap \ker d$, and so $x = 0$. A similar argument shows that $\varphi : M_{k+1} \to M_{k+1}'$ is surjective.

$\square$

5.3 Acyclic Closures and $\overline{UL}$

Recall [36, Chap 3] that the *acyclic closure* is the special case, $\wedge V \otimes \wedge U$, of a Λ-extension constructed inductively as follows. Write $V = \cup_n V_n$ with $V_0 = V \cap \ker d$ and $V_{n+1} = V \cap d^{-1}(\wedge^2 V_n)$. Then $U = \cup_{n \geq 0} U_n$, and there is a degree 1 isomorphism $p : U \xrightarrow{\cong} V$ restricting to isomorphisms $U_n \xrightarrow{\cong} V_n$. Thus this identifies $L = U^{\vee}$. Finally, it follows from [36, Proposition 3.12] that the differential is determined by the conditions

$$du = pu, \quad u \in U_0 \quad \text{and} \ (d - p) : U_{n+1} \to V_n \otimes \wedge U_n.$$

In particular the augmentation $\varepsilon_U : \wedge U \to \mathbb{Q}, U \to 0$, together with the unique augmentation in $\wedge V$ defines a quasi-isomorphism

$$\wedge V \otimes \wedge U \xrightarrow{\cong} \mathbb{Q}.$$

This construction identifies $\wedge V \otimes \wedge U$ as a quadratic $(\wedge V, d)$-module. In this case the holonomy representation is a representation $\theta : L \to \mathrm{Der}(\wedge U)$ of L by derivations in $\wedge U$

$$\theta(x) \cdot u = -\sum \langle v_i, sx \rangle \alpha_i,$$

where $d(1 \otimes u) = \sum v_i \otimes \alpha_i, \alpha_i \in \wedge U$.

Therefore if $x \in L$ and $u \in U_n$, then

$$\theta(x)u + \ < pu, sx > \in V_{n-1} \otimes \wedge^{\geq 1} U_{n-1}. \tag{5.3}$$

Now right multiplication by L in $\overline{UL}$ makes (by definition) $\overline{UL}$ into a right enriched L-module. On the other hand the holonomy representation in $\wedge U$ dualizes to make $(\wedge U)^{\vee}$ into a right enriched L-module and hence into a $\overline{UL}$-module.

Proposition 5.2 *An isomorphism*

$$\eta_L : \overline{UL} \xrightarrow{\cong} (\wedge U)^{\vee}$$

of right L-modules is defined by

$$< \Phi, \eta_L(a) > = (-1)^{\deg a} \varepsilon_U(a \cdot \Phi), \quad a \in \overline{UL}, \Phi \in \wedge U.$$

Proof First observe that for $x \in L$,

$$\begin{aligned} < x \cdot \Phi, \eta_L(a) > &= (-1)^{\deg a} \varepsilon_U(a \cdot x \cdot \Phi) \\ &= (-1)^{\deg x} < \Phi, \eta_L(ax) > . \end{aligned}$$

Thus η_L is a morphism of right L-modules. In [36, Theorem 6] it is proved that η_L is an isomorphism if $\dim L_0/[L_0, L_0] < \infty$ and each L_k is finite dimensional. Thus each η_{L_α} is an isomorphism. It follows that $\eta_L = \varprojlim_\alpha \eta_{L_\alpha}$ is an isomorphism. $\qquad\square$

5.4   Profree *L*-Modules

Definition Let L be an enriched Lie algebra and V an enriched vector space. The *profree L-module freely generated by V* is the L-module

$$V \widehat{\otimes} \overline{UL}.$$

Profree L-modules satisfy the following property: If M is an enriched L-module, every morphism of enriched vector spaces $f : V \to M$ extends in a unique way to a map of enriched L-module $V \widehat{\otimes} \overline{UL} \to M$.

In particular a morphism of enriched L-module of $M \widehat{\otimes} \overline{UL}$ to itself is the identity if it induces the identity on M.

If $\wedge V \otimes \wedge U$ is the acyclic closure of the quadratic Sullivan model of L, then $(\wedge U)^\vee \cong \overline{UL}$ is the profree module freely generated by $\mathbb{Q}$ (Proposition 5.2).

5.5   Closure of a Subspace for the Adjoint Action

Lemma 5.2 *If $S \subset L$ is any subspace, then for the adjoint action, $S \cdot \overline{UL}$ is an ideal. If S is a graded space of finite type, then $S \cdot \overline{UL}$ is closed.*

Proof Observe that

$$[x \cdot \Phi, y] = x \cdot \Phi y, \quad x, y \in L, \Phi \in \overline{UL},$$

and so $S \; \overline{UL}$ is an ideal. Since a graded space is closed if and only if each component in a given degree is closed, if S has finite type, it is sufficient to show that each $S^k \cdot \overline{UL}$ is closed. Since a finite sum of closed subspaces is closed, we have only to show that $x \cdot \overline{UL}$ is closed for each $x \in L$.

Recall that $\overline{UL} = \varprojlim_\alpha \widehat{UL_\alpha}$, and denote by J_α the augmentation ideal of UL_α with completion $\widehat{J_\alpha}$. Then $\overline{UL} = \mathbb{Q} \oplus \overline{J}$, with $\overline{J} = \varprojlim_\alpha \widehat{J_\alpha}$.

Denote $\rho_\alpha x$ by x_α. Then the closure of $x \cdot \overline{UL}$ is the inverse limit

$$\varprojlim_\alpha \rho_\alpha(x \cdot \overline{UL}) = \varprojlim_\alpha x_\alpha \cdot \widehat{UL_\alpha} = \mathbb{Q}x \oplus \varprojlim_\alpha x_\alpha \cdot \widehat{J_\alpha} = \mathbb{Q}x \oplus \varprojlim_{n,\alpha}(x_\alpha \cdot J_\alpha)/(x_\alpha \cdot J_\alpha^n).$$

On the other hand, because each J_α/J_α^n is finite dimensional, the surjections $J_\alpha/J_\alpha^n \to (x_\alpha \cdot J_\alpha)/x_\alpha \cdot J_\alpha^n)$ induce a surjection

$$\overline{J} \to \varprojlim_{n,\alpha} x_\alpha \cdot J_\alpha/(x_\alpha \cdot J_\alpha^n).$$

This factors through the map $\overline{J} \to x \cdot \overline{J}$ and therefore shows that

$$x \cdot \overline{J} \to \varprojlim_\alpha \rho_\alpha(x \cdot \overline{J})$$

is surjective. But this is the inclusion of $x \cdot \overline{J}$ in its closure, and so $x \cdot \overline{J}$ is closed.

$\qquad\square$

5.6 The L/I-Module $I/I^{(2)}$

Recall from Sect. 4.1 that a closed ideal $I \subset L$ decomposes $\wedge V$ as a Λ-extension $\wedge V = \wedge W \otimes \wedge Z$ in which the inclusion $(\wedge W, d_W) \to (\wedge V, d)$ and the surjection $(\wedge V, d) \to (\wedge Z, d_Z)$ are morphisms of quadratic Sullivan algebras corresponding, respectively, to the morphisms

$$L \to L/I = L_W \quad \text{and} \quad I = L_Z \hookrightarrow L.$$

Thus filtering by the ideals $\wedge^{\geq k} W \otimes \wedge Z$ produces a spectral sequence converging from $(\wedge W \otimes H(\wedge Z, d_Z), \overline{d})$ to $H(\wedge V, d)$. Note that since $(\wedge Z, d_Z)$ is quadratic,

$$H(\wedge Z, d_Z) = \oplus_k H^{[k]}(\wedge Z, d_Z),$$

where $H^{[k]}(\wedge Z)$ is the subspace represented by classes having a representative in $\wedge^k Z$. Moreover, by definition

$$\overline{d} : 1 \otimes H^{[k]}(\wedge Z) \to W \otimes H^{[k]}(\wedge Z).$$

It follows that each $(\wedge W \otimes H^{[k]}(\wedge Z, \overline{d})$ is a quadratic $(\wedge W, d)$-module. Hence this defines a holonomy representation of $L_W = L/I$ in each $H^{[k]}(\wedge Z)$ (Sect. 5.3).

Proposition 5.3 *With the hypotheses and notation above, there is a natural isomorphism*

$$s^{-1}(Z \cap ker\, d_Z)^{\vee} = s^{-1} H^{[1]}(\wedge Z)^{\vee} \xrightarrow{\cong} I/I^{(2)}$$

of enriched L/I-modules, in which:

- *L/I acts in $H^{[1]}(\wedge Z)^{\vee}$ by the dual of the holonomy representation.*
- *L/I acts in $I/I^{(2)}$ by the representation induced from the right adjoint representation of L in I.*

Proof By definition, $H^{[1]}(\wedge Z, d_Z) = Z \cap ker\, d_Z$, while from Lemma 4.1 we obtain

$$s^{-1}(Z \cap ker\, d_Z)^{\vee} = L_Z/L_Z^{(2)} = I/I^{(2)}.$$

It remains to verify that this isomorphism is an isomorphism of L/I-modules.

Denote the holonomy representation of L/I in $Z \cap ker\, d_Z$ by θ. Then if $z \in Z \cap ker\, d_Z$,

$$\overline{d}z = \sum w_i \otimes z_i$$

with $w_i \in W$ and $z_i \in Z \cap ker\, d_Z$. Thus, for $y \in L$, $\theta(y)$ is determined by the equation (Sect. 5.2)

$$< \theta(y)z, sx > = -\sum_i < w_i, sy > < z_i, sx > = - < dz, sx, sy > .$$

Now it follows from Proposition 5.1(iv) that

$$< \theta(y)z, sx > = < z, ad^r x(y) > .$$

$\square$

Profree Lie Algebras

6

Profree Lie algebras are the analogs of the free Lie algebras in the category of enriched Lie algebras. If $V = \varprojlim V_\alpha$, the profree Lie algebra on V, $\overline{\mathbb{L}}_V$, is the inverse limit of the $\overline{\mathbb{L}}_{V_\alpha}$, where $\overline{\mathbb{L}}_{V_\alpha}$ is the pronilpotent completion of the free Lie algebra $\mathbb{L}_{V_\alpha}$. The main point of the section is a characterization of the profree Lie algebras in terms of the associated quadratic Sullivan algebras.

6.1   Definition of a Profree Lie Algebra

Free graded Lie algebras play a key role in Lie algebra theory. Profree Lie algebras are the analogs of free Lie algebras in the category of enriched Lie algebras. We denote by $\mathbb{L}_T$ the usual free graded Lie algebra generated by the graded vector space T.

Definition A *profree Lie algebra*, L, is an enriched Lie algebra of the form $L \cong \overline{\mathbb{L}}_T$ in which $(T, \{\rho_\alpha\})$ is an enriched vector space and the enriched structure in $\mathbb{L}_T$ is defined by the surjections

$$\mathbb{L}_T \to \mathbb{L}_{\rho_\alpha T} \to \mathbb{L}_{\rho_\alpha T}/\mathbb{L}^n_{\rho_\alpha T}.$$

We call $\overline{\mathbb{L}}_T$ the profree Lie algebra generated by the enriched vector space T.

Remark There is clearly a natural injective map $\mathbb{L}_T \to \overline{\mathbb{L}}_T$.

Proposition 6.1 *Let $(T, \{\rho_\alpha\})$ be an enriched vector space.*

Y. Félix, S. Halperin, *Lie Models for Spaces*, Frontiers in Mathematics,
https://doi.org/10.1007/978-3-032-15357-9_6

(i) *Any morphism $f : T \to E$ into an enriched Lie algebra E extends uniquely to a morphism*

$$\overline{\mathbb{L}}_T \to E$$

of enriched Lie algebras.

(ii) $\overline{\mathbb{L}}_T = \overline{\mathbb{L}}_T^{(2)} \oplus T$, *and $T \subset \overline{\mathbb{L}}_T$ is closed. Moreover, if $L \cong \overline{\mathbb{L}}_T$ is profree, then division by $L^{(2)}$ induces an isomorphism $L/L^{(2)} \xrightarrow{\cong} T$ of enriched vector spaces.*

(iii) *If T' is any closed direct summand of $\overline{\mathbb{L}}_T^{(2)}$ in $\overline{\mathbb{L}}_T$, then the inclusion extends to an isomorphism*

$$\overline{\mathbb{L}}_{T'} \xrightarrow{\cong} \overline{\mathbb{L}}_T$$

of enriched Lie algebras.

(iv) *If T is a graded vector space of finite type, then the completion of the universal enveloping algebra $U\overline{\mathbb{L}}_T$ is given by*

$$\overline{U\overline{\mathbb{L}}_T} = \prod_{n \geq 0} T^{\otimes^n}$$

where $\otimes^n$ denotes the nth tensor power.

Proof

(i) Denote by $E \to E_\beta$ the surjections in the enriched structure of E. Since E_β is finite dimensional and nilpotent, it follows that the extension of f to a morphism $\mathbb{L}_T \to E$ factors to yield morphisms

$$\mathbb{L}_{\rho_{\alpha(\beta)} T} \to E_\beta.$$

Passing to inverse limits gives the morphism $\overline{\mathbb{L}}_T \to E$, and it is immediate that it is the unique extension of f.

(ii) Since $\mathbb{L}_{\rho_\alpha T} = \rho_\alpha T \oplus \mathbb{L}^2_{\rho_\alpha T}$, it follows from Lemma 2.2 that

$$\overline{\mathbb{L}}_T^{(2)} = \varprojlim_\alpha \mathbb{L}_{\rho_\alpha} T/\rho_\alpha T = \overline{\mathbb{L}}_T/T.$$

This gives (ii).

(iii) If T' is a second closed direct summand of $\overline{\mathbb{L}}_T^{(2)}$ in $\overline{\mathbb{L}}_T$, then (ii) provides morphisms

$$\alpha : \overline{\mathbb{L}}_{T'} \to \overline{\mathbb{L}}_T \quad \text{and} \quad \beta : \overline{\mathbb{L}}_T \to \overline{\mathbb{L}}_{T'}$$

which induce isomorphisms $\overline{\mathbb{L}}_{T'}/\overline{\mathbb{L}}_{T'}^{(2)} \cong \overline{\mathbb{L}}_{T}/\overline{\mathbb{L}}_{T}^{(2)}$. Now filtering by the ideals $\overline{\mathbb{L}}_{T'}^{(n)}$ and $\overline{\mathbb{L}}_{T}^{(n)}$ shows that α and β are inverse isomorphisms (Lemma 3.1).

(iv) Because T has finite type, (cf. Sect. 1.2)

$$\overline{U\overline{\mathbb{L}}_T} = \widehat{U\overline{\mathbb{L}}_T} = \widehat{\oplus_n T^{\otimes^n}} = \prod_n T^{\otimes^n}$$

where $\oplus\, T^{\otimes^n}$ is the tensor algebra on T. $\square$

6.2   Theorem 6.1: Sullivan Characterization of a Profree Lie Algebra

Theorem 6.1 *The following conditions are equivalent on an enriched Lie algebra, L:*

(i) $L = \overline{\mathbb{L}}_T$ *is profree.*
(ii) The quadratic model $\wedge V$ *of L satisfies* $H(\wedge V) = \mathbb{Q} \oplus (V \cap \ker d)$.

In this case, $sT \xrightarrow{\cong} (V \cap \ker d)^{\vee}$.

Before entering in the proof we need complementary properties of quadratic Sullivan algebras.

6.2.1   Quadratic Sullivan Algebras

For any quadratic Sullivan algebra, $\wedge V$, recall (Chap. 4) that the standard filtration of V is given by

$$V_0 = V \cap \ker d \quad \text{and} \quad V_{n+1} = V \cap d^{-1}(\wedge^2 V_n).$$

In particular,

$$H(\wedge V) = \oplus_k H^{[k]}(\wedge V),$$

where $H^{[k]}(\wedge V)$ is the image in $H(\wedge V)$ of the space of cycles in $\wedge^k V$. Our objective here is to prove (Lemma 6.2) that

$$H^{[2]}(\wedge V) = 0 \iff H(\wedge V) = \mathbb{Q} \oplus (V \cap \ker d).$$

Lemma 6.1 *Suppose* $\psi : E \to L$ *is a morphism of enriched Lie algebras.*

(i) ψ *is surjective if and only if* $\psi(2) : E/E^{(2)} \to L/L^{(2)}$ *is surjective.*

(ii) Suppose the quadratic model, $\wedge V$, of L satisfies $H^{[2]}(\wedge V) = 0$. Then the following assertions are equivalent:
   (a) $\psi(2)$ is an isomorphism.
   (b) $\psi(n)$ is an isomorphism, $n \geq 2$.
   (c) ψ is an isomorphism.

Proof First recall that ψ is the dual of $\varphi : V \to W$ where $\varphi : \wedge V \to \wedge W$ is the corresponding morphism between the quadratic models of L and E. In view of Lemma 4.1, $\psi(n + 2) : E/E^{(n+2)} \to L/L^{(n+2)}$ is the dual of $\varphi_n : V_n \to W_n$.

(i) We have only to show that if φ_0 is injective, then φ is injective. Assume by induction that φ_n is injective. If $v \in V_{n+1}$ and $\varphi v = 0$, then since $dv \in \wedge^2 V_n$ and φ_n is injective it follows that $dv = 0$. Thus $v \in V_0$, and since φ_0 is injective $v = 0$.

(ii) Suppose (a) is satisfied, and assume by induction that $\varphi_n : V_n \overset{\cong}{\to} W_n$. If $w \in W_{n+1}$, then dw is a cycle in $\wedge^2 W_n$, and so $dw = \varphi\Phi$, where Φ is a cycle in $\wedge^2 V_n$. Because $H^{[2]}(\wedge V) = 0$, it follows that $\Phi = dv$ for some $v \in V_{n+1}$. Thus $d(w - \varphi v) = 0$ and so $w - \varphi(v) \in W_0$. By hypothesis, $w - \varphi v = \varphi v_0$ for some $v_0 \in V_0$, and so $W_{n+1} \subset \varphi(V_{n+1})$. On the other hand, by (i), $\varphi_{n+1} : V_{n+1} \to W_{n+1}$ is injective. This proves that (a) $\Rightarrow$ (b).

But then $\varphi = \varinjlim \varphi_n$ and so φ is an isomorphism. Thus (b) $\Rightarrow$ (c). Finally, it is immediate that if φ is an isomorphism, then so is φ_0. $\qquad\square$

Lemma 6.2 *If a morphism $\rho : \wedge V \to \wedge Z$ of quadratic Sullivan algebras restricts to a surjection $\rho_0 : V_0 \to Z_0$ and if $H^{[2]}(\wedge V) = 0$, then*

$$H^{[k]}(\wedge Z) = 0, \quad k \geq 2.$$

Proof The proof is in two steps.
Step 1. *If, in addition, ρ is surjective, then $H^{[k]}(\wedge Z) = 0, k \geq 2$.*
   Here we first show that $H^{[2]}(\wedge Z) = 0$. Since ρ is surjective, by [36, Cor.3.4] ρ extends to a quasi-isomorphism

$$\rho : \wedge V \otimes \wedge U \overset{\simeq}{\to} \wedge Z$$

in which $\wedge V \otimes \wedge U$ is a Λ-extension of $\wedge V$ and $d : \wedge U \to V \otimes \wedge U$.

   More precisely $\wedge V \otimes \wedge U$ is constructed by induction on the standard filtration of $\wedge V$. We define $\rho : \wedge V \otimes \wedge U \to \wedge Z$ as the union of the morphisms $\rho_n = \wedge V_n \otimes \wedge U_n \to \wedge Z_n$ satisfying the following conditions:

- $d(U_n) \subset V_n + (V_{n-1} \otimes \wedge U_{n-1})$.
- $\rho_n(U_n) = 0$.
- ρ_n is a quasi-isomorphism.

Suppose ρ_n has been constructed, then let K be the kernel of $\rho : V_{n+1} \to Z_{n+1}$, and let $p : U_{n+1} \to K$ be an isomorphism of degree 1. Then for $x \in U_{n+1}$, dpx is a cycle and $\rho dp(x) = 0$. Therefore $dpx = d\alpha$, with $\alpha \in V_n \otimes \wedge^{\geq 1} U_n$. We define $dx = px - \alpha$, and ρ_{n+1} is clearly a quasi-isomorphism.

Now, putting V and Z in gradation 1 and U in gradation zero, ρ is an isomorphism preserving the gradation. Therefore there is a quasi-isomorphism $\sigma : \wedge Z \to \wedge V \otimes \wedge U$ satisfying $\rho \circ \sigma = id$ and $\sigma(Z) \subset V \otimes \wedge U$. In particular, the decomposition $\wedge V \otimes \wedge U = \oplus_k \wedge^k V \otimes \wedge U$ induces a decomposition of its homology, and

$$\sigma : H^{[k]}(\wedge Z) \xrightarrow{\cong} H^{[k]}(\wedge V \otimes \wedge U), \qquad k \geq 0.$$

On the other hand, by construction, $\wedge U$ is the increasing union of the subspaces $(\wedge U)_q$ and

$$H^{[k]}(\wedge V \otimes \wedge U) = \varinjlim_q H^{[k]}(\wedge V \otimes (\wedge U)_q).$$

But the differential in the quotients $\wedge V \otimes ((\wedge U)_{q+1}/(\wedge U)_q)$ is just $d \otimes id$, and it follows that $H^{[2]}(\wedge V \otimes \wedge U) = 0$.

It remains to show that $H^{[k]}(\wedge Z) = 0$, $k \geq 3$. Suppose $\Phi \in \wedge^k Z$ is a cycle. There is then a sequence $z_1, \ldots, z_r$ of elements in Z such that $dz_1 = 0$, $dz_{i+1} \in \wedge^2(z_1, \ldots, z_i)$, $\Phi \in \wedge(z_1, \ldots, z_r)$ and such that division by $z_1, \ldots, z_r$ maps Φ to zero. When $r = 1$, $\Phi = z_1^k$ is a boundary. We use induction on r and on k to show that Φ is a boundary.

Observe first that division by z_1 gives a quadratic Sullivan algebra $(\wedge Z', d')$. By what we just proved, $H^{[2]}(\wedge Z', d') = 0$. Thus by induction on r, the image of Φ in $\wedge Z'$ is a boundary. It follows that for some $\Phi' \in \wedge^{k-1} Z$,

$$\Phi - d(1 \otimes \Phi') = z_1 \otimes \Phi'',$$

with $\Phi'' \in \wedge Z$. In particular, in $\wedge Z'$, Φ'' is a d'-cycle. By induction on k in $\wedge Z'$, $\Phi'' = d'\Omega$ for some $\Omega \in \wedge^{k-2} Z'$. Therefore $\Phi'' = d\Omega + z_1 \otimes \Psi$ for some Ψ.

If $\deg z_1$ is odd, then $z_1 \otimes \Phi'' = d(-z_1 \otimes \Omega)$, and so Φ is a boundary. If $\deg z_1$ is even, then, since $H^{[2]}(\wedge Z) = 0$, we may choose z_2, so $dz_2 = z_1^2$. Division by z_2 and z_1^2 then gives a quasi-isomorphism $\wedge Z \xrightarrow{\cong} ((\wedge z_1)/z_1^2) \otimes \wedge Z''$, and the same argument as above shows that Φ is a boundary.

Step 2. *Completion of the proof of Lemma 6.2*

We define a sequence of surjective morphisms

$$\wedge V = \wedge V(1) \twoheadrightarrow \wedge V(2) \twoheadrightarrow \cdots \twoheadrightarrow \wedge V(p) \twoheadrightarrow \cdots$$

such that ρ factors through each to yield morphisms

$$\rho(p) : \wedge V(p) \twoheadrightarrow \wedge Z.$$

In fact, if $\rho(p)$ has been defined, let $\varphi_p : \wedge V(p) \twoheadrightarrow \wedge V(p+1)$ be obtained by division by $V(p)_0 \cap \ker\rho(p)$.

Then the kernels of the surjections $\wedge V \to \wedge V(p)$ form an increasing sequence of subspaces $K(p) \subset V$. Set $K = \cup_p K(p)$, and let

$$\varphi : \wedge V \to \wedge W$$

be the surjection obtained by dividing V by K. Then $\varphi = \varinjlim_p \varphi_p$, and so φ and φ_0 are surjective. Thus by Step 1 $H^{[k]}(\wedge W) = 0, k \geq 2$.

Now by construction, ρ factors as

$$\wedge V \xrightarrow{\varphi} \wedge W \xrightarrow{\gamma} \wedge Z.$$

Moreover, $\wedge W = \varinjlim_p \wedge V(p)$ and so $W_0 = \varinjlim_p V(p)_0$. Let $I(p)$ be the image of $V(p)_0$ in $V(p+1)_0$. Then also

$$W_0 = \varinjlim_p I(p).$$

Thus by construction, $\gamma : W_0 \xrightarrow{\cong} Z_0$. Now Lemma 6.1(ii) asserts that γ is an isomorphism, and hence

$$H^{[k]}(\wedge Z) = H^{[k]}(\wedge W) = 0, \quad k \geq 2.$$

$\square$

6.2.2 Proof of Theorem 6.1

This is in two steps. We fix a closed summand, T, of $L^{(2)}$ in L, and we denote by $(\wedge V, d)$ the quadratic model of L. Then it follows from Lemma 4.1 that the isomorphism $L \xrightarrow{\cong} V^\vee$ restricts to an isomorphism $T \xrightarrow{\cong} (V \cap \ker d)^\vee$.

Step 1. If $L = \overline{\mathbb{L}}_T$ is profree, then $H(\wedge V) = \mathbb{Q} \oplus (V \cap \ker d)$.

In this case, by adjoining additional variables construct an inclusion $\lambda : \wedge V \to \wedge W$ of quadratic Sullivan algebras for which

$$V_0 \overset{\cong}{\longrightarrow} W_0 \quad \text{and} \quad H^{[2]}(\wedge W) = 0.$$

Dualizing $V \to W$ gives a surjection $L \overset{\rho}{\leftarrow} L_W$ of enriched Lie algebras. Moreover, since $V_0 \overset{\cong}{\to} W_0$, $L_W = L_W^{(2)} \oplus T_W$, where T_W is closed and

$$\rho_T : T \overset{\cong}{\longleftarrow} T_W .$$

Here both ρ_T and its inverse, σ, are morphisms of enriched vector spaces and so σ extends to a morphism $\varphi : L \to L_W$. Since $\rho_T \circ \sigma = id_T$, it follows that $\rho \circ \varphi = id_L$. Now dualizing, φ, gives a morphism

$$\psi : \wedge V \leftarrow \wedge W$$

such that $\psi \circ \lambda = id_{\wedge V}$.

On the other hand, the inclusion λ defines a Λ-extension

$$\wedge V \otimes \wedge Z \overset{\cong}{\longrightarrow} \wedge W,$$

in which if $Z \neq 0$, then some $z \in Z$ satisfies $dz \in \wedge^2 V$. Thus

$$d\psi z = \psi dz = dz.$$

Hence $\psi z - z$ is a cycle in W. This gives

$$\psi z - z \in W_0 = V_0$$

and so $z \in V$. This contradicts $z \in Z$, and it follows that $Z = 0$ and ψ is an isomorphism inverse to λ. In particular $H^{[2]}(\wedge V) = 0$, and now Lemma 6.2 implies that $H(\wedge V) = \mathbb{Q} \oplus (V \cap \ker d)$.

Step 2. If $H(\wedge V) = \mathbb{Q} \oplus (V \cap \ker d)$, then L is profree.

In this case, by Proposition 6.1, the inclusion $T \to L$ extends to a morphism

$$\varphi : \overline{\mathbb{L}}_T \to L.$$

On the one hand, Proposition 4.2 implies that the image of φ is closed. On the other hand, since the image contains T, Lemma 3.2 implies that φ is surjective. And, finally, Proposition 4.2 also asserts that $\ker \varphi$ is closed.

Now, taking the quadratic models for the short exact sequence (Sect. 4.1)

$$0 \leftarrow L \leftarrow \overline{\mathbb{L}}_T \leftarrow \ker \varphi \leftarrow 0$$

yields the Λ-extension

$$\wedge V \to \wedge V \otimes \wedge Z \to \wedge Z$$

in which $\wedge V \otimes \wedge Z$ is the quadratic model of $\overline{\mathbb{L}}_T$. We show that φ is an isomorphism by proving that $Z = 0$.

But if $Z \neq 0$, then for some $z \neq 0$ in Z, $dz \in \wedge^2 V$. Since by hypothesis $H^{[2]}(\wedge V) = 0$, this implies that $dz = dv$ some $v \in V$. But then $v - z$ is a cycle representing a class in $H^{[1]}(\wedge V \otimes \wedge Z)$. However, because φ induces an isomorphism $\overline{\mathbb{L}}_T/\overline{\mathbb{L}}_T^{(2)} \to L/L^{(2)}$, it follows that

$$V \cap \ker d = H^{[1]}(\wedge V) \overset{\cong}{\to} H^{[1]}(\wedge V \otimes \wedge Z) = (V \oplus Z) \cap \ker d.$$

Thus $z = 0$, and therefore $Z = 0$.

The final isomorphism $sT \cong (V \cap \ker d)^{\vee}$ is a particular case of Lemma 4.1. $\square$

6.3   Properties of Profree Lie Algebras

We have the following from Theorem 6.1:

Corollary 6.1 *Suppose $E = R \oplus E^2$ is a free graded Lie algebra. If $\dim R < \infty$ and E is equipped with its unique (Proposition 3.3) enriched structure, then $\overline{E}$ is profree.*

Proof By Proposition 3.1, $\overline{E} = R \oplus \overline{E}^{(2)}$, and by definition $\overline{E} = \overline{\mathbb{L}}_R$. $\square$

Corollary 6.2 *If E is an enriched Lie algebra, then the inclusion of a direct summand T of $E^{(2)}$ in E extends to a surjection $\overline{\mathbb{L}}_T \to E$.*

Proof The existence of the morphism follows from Proposition 6.1. Since its image is a closed sub-Lie algebra (Proposition 2.1) and contains T, the morphism is surjective. $\square$

Corollary 6.3 *If $\psi : (L, \mathcal{I}) \to (F, \mathcal{G})$ is a surjective morphism from an enriched Lie algebra to a profree Lie algebra, then there is a morphism*

$$\sigma : (F, \mathcal{G}) \to (L, \mathcal{I})$$

such that $\psi \circ \sigma = id_F$.

Proof Write $F = \overline{\mathbb{L}}_T$. The surjection $\overline{\mathbb{L}}_T \to T$ composed with ψ yields a surjection $L \to T$. By Proposition 2.1 this admits a splitting, $T \to L$. This extends to a morphism $\sigma : \overline{\mathbb{L}}_T \to L$, and since $\psi \circ \sigma_{|T} = id_T$, it follows that $\psi \circ \sigma = id$. $\square$

Now recall from [36, Chap 9] that the Lusternik-Schnirelman category of a minimal Sullivan algebra, $\wedge V$, is the least p (or ∞) such that $\wedge V$ is a homotopy retract of $\wedge V / \wedge^{>p} V$.

Proposition 6.2

(i) *An enriched Lie algebra is profree if and only if its quadratic model satisfies* $cat(\wedge V) = 1$.

(ii) *Any closed sub-Lie algebra of a profree Lie algebra is profree.*

Proof

(i) The condition $cat(\wedge V) = 1$ for a quadratic Sullivan algebra is equivalent to the condition $\mathbb{Q} \oplus (V \cap \ker d) \xrightarrow{\cong} \wedge V$. Thus (i) follows from Theorem 6.1.

(ii) Suppose $\wedge V \to \wedge Z$ is the morphism of quadratic models corresponding to an inclusion $E \to L$ of a closed sub-Lie algebra in an enriched Lie algebra. Then [36, Theorem 9.3] gives $cat(\wedge Z) \leq cat(\wedge V)$. Thus (ii) follows from (i). $\square$

Corollary *A solvable sub-Lie algebra, E, of a profree Lie algebra, L, is either 0 or the free Lie algebra on a single generator.*

Proof Suppose $E \neq 0$. By Proposition 6.2, $\overline{E} = \overline{\mathbb{L}}_S$ for some closed subspace $S \subset \overline{E}$. Lemma 3.2 implies that $\overline{\mathbb{L}}_S$ is solvable.

On the other hand, denote by $L^{[r]}$ the derived series of L

$$L^{[0]} = L, \quad \text{and } L^{[r+1]} = [L^{[r]}, L^{[r]}].$$

By Lemma 3.2 (ii) we have $\overline{\mathbb{L}_S^{[r]}} = \overline{\mathbb{L}_S}^{[r]}$, and hence $\mathbb{L}_S$ is solvable. It follows that $\mathbb{L}_S$ is a free Lie algebra on a single generator. In particular $\mathbb{L}_S$ has finite type. This in turn implies that $\mathbb{L}_S = \overline{\mathbb{L}}_S = \overline{E} = E$. $\square$

Remark Recall that any sub-Lie algebra of a free Lie algebra is free [75]. By Proposition 6.2(ii) the analogous statement for closed sub-Lie algebras of profree Lie algebras is also true.

Example Let $L = \overline{\mathbb{L}}_T$ be a profree Lie algebra. Then the sub-Lie algebra generated by T is free. The closed sub-Lie algebras of L are profree, but a sub-Lie algebra is not necessarily free. For instance let $L = \widehat{\mathbb{L}(a, b)}$ and $E \subset L$ be the sub-Lie algebra generated by the elements a, b, and $\omega = \sum_{n \geq 1} \mathrm{ad}_a^n(b)$. Since $\omega = [a, \omega] - [a, b]$, $\omega \in E^2$, so that E/E^2 is generated by the classes of a and b. If E was free, then E would be equal to $\mathbb{L}(a, b)$, which is impossible because $\omega \notin \mathbb{L}(a, b)$.

6.4   Inverse Limits of Profree Lie Algebras

Proposition 6.3 *Let L be an enriched Lie algebra.*

(i) *Suppose $L/[L, L]$ has finite type. Then L is profree if and only if the inclusion $T \to L$ of a direct summand of $[L, L]$ extends to an isomorphism*

$$\widehat{\mathbb{L}_T} := \varprojlim_n \mathbb{L}_T /\mathbb{L}_T^n \xrightarrow{\cong} L.$$

(ii) *If $L = \varprojlim_\sigma L_\sigma$ is any inverse limit of profree Lie algebras, then L is profree.*

Proof

(i) Proposition 3.2 identifies $\widehat{\mathbb{L}_T} \xrightarrow{\cong} \overline{\mathbb{L}}_T$, since $\mathbb{L}_T$ has a unique enriched structure. Thus (i) follows from Theorem 6.1.

(ii) If $\wedge V$ and $\wedge V_\sigma$ are, respectively, the quadratic models of L and L_σ, then $\wedge V = \varinjlim_\sigma \wedge V_\sigma$, and the map $V \cap \ker d \to \varinjlim_\sigma V_\sigma \cap \ker d$ is an isomorphism. It follows (since each L_σ is profree) from Theorem 6.1 that $V \cap \ker d \xrightarrow{\cong} H^{\geq 1}(\wedge V)$ and hence that L is profree. $\qquad\square$

Proposition 6.4 *Let $L = \varprojlim L_\sigma$ be the inverse limit of a family of finitely generated profree Lie algebras. Then for each $k \geq 1$,*

$$L^{(k)} = \varprojlim_\sigma L(\sigma)^k.$$

Proof Since $L = \varprojlim_\sigma L(\sigma)$, it follows that $V = \varinjlim_\sigma V_\sigma$ where $\wedge V_\sigma$ is the quadratic model of $L(\sigma)$. Recall from Chap. 4 the standard filtration $V = \varinjlim_k V_k$. Then

$$V/V_k = \varinjlim_\sigma V(\sigma)/V(\sigma)_k,$$

and this dualizes to $L^{(k)} = \varprojlim_\ell L(\sigma)^{(k)}$ (Lemma 4.1). Since $L(\sigma) = \widehat{\mathbb{L}}_{T(\omega)}$, we have $L(\sigma)^{(k)} = L(\sigma)^k$.

$$\qquad\square$$

6.5   Relation Between L^2 and $L^{(2)}$

When L is finitely generated, $L^k \cong L^{(k)}$. This is not the case in general.

Proposition 6.5 *If $L = L_0$ is a profree Lie algebra and $\dim L/[L, L] = \infty$, then $L^2 \neq L^{(2)}$. If, in particular, $L/L^{(2)}$ is concentrated in degree r, then $L_{2r}^2 \neq L_{2r}^{(2)}$.*

Proof Indeed, let $(\wedge W, d)$ be the quadratic model of L. We can decompose W as a union $W = \cup_n W_n$ with $W_0 = W \cap \ker d$, and for $n > 1$, $W_n = d^{-1}(\wedge^2 W_{n-1})$. We denote by Z_n a direct summand of W_{n-1} in W_n. Then by hypothesis W_0 is infinite dimensional, and there is a quasi-isomorphism $\varphi : (\wedge W, d) \to (\mathbb{Q} \oplus W_0, 0)$ that is the identity on W_0 and that maps each W_n to 0 for $n \geq 1$.

For the sake of simplicity, we write $V = W_0$ and $Z = W_1$. By construction the isomorphism $sL \xrightarrow{\cong} W^\vee$ induces isomorphisms

$$s(L/L^{(2)}) \cong V^\vee, \qquad \text{and } s(L^{(2)}/L^{(3)}) \cong Z^\vee.$$

Suppose first that W is infinite countable, and denote by $w_1, w_2, \ldots$ a basis of V. Since $d = 0$ on V, $d : Z \to \wedge^2 V$ is an isomorphism. We denote by w_{ij}, $i < j$ the basis of Z defined by $d(w_{ij}) = w_i \wedge w_j$. Denote by E the vector space of column matrices with rational numbers $X = (a_i)$ with only a finite number of nonzero elements a_i. Then the map $(a_i) \rightsquigarrow \sum a_i w_i$ defines an isomorphism $E \xrightarrow{\cong} V$.

Let us represent an element $\varphi \in Z^\vee$ by the infinite dimensional antisymmetric matrix M_φ defined by

$$(M_\varphi)_{ij} = \varphi(w_{ij}), \quad i < j.$$

By a similar process, an element $f \in V^\vee$ can be represented by a column matrix A_f, with $(A_f)_i = f(w_i)$. The vector space $\ker f$ can then be identified with the subvector space of E formed by the column matrices X satisfying $A_f^t \cdot X = 0$. (Here A^t denotes the line matrix transposition of the column matrix A.)

Note that when we have two column matrices A and B, we can form the antisymmetric matrix $A \cdot B^t - B \cdot A^t$. Now remark that for $f, g \in V^\vee$, we have

$$M_{[f,g]} = A_f \cdot B_g^t - B_g \cdot A_f^t.$$

Let $\varphi_0 \in Z^\vee$ be the particular element defined for $i < j$ by

$$\varphi_0(w_{ij}) = \begin{cases} 1 \text{ if } w_{ij} = w_{2k+1,2k+2}, \text{ for some } k \\ 0 \text{ otherwise} \end{cases}$$

The element φ_0 can be extended to all of W, by $\varphi_0(V) = 0$ and $\varphi_0(Z_n) = 0$, for $n > 1$. By construction $\varphi_0 \in L^{(2)}$. The associated matrix is

$$M_0 = \begin{pmatrix} B_0 & 0 & 0 & \dots \\ 0 & B_0 & 0 & \dots \\ 0 & 0 & B_0 & \dots \\ & \dots\dots\dots & \end{pmatrix} \qquad \text{with } B_0 = \begin{pmatrix} 0 & 1 \\ -1 & 0 \end{pmatrix}.$$

Now consider a finite sum $\sum_{i=1}^{n} [f_i, g_i]$, with f_i and $g_i \in V^\vee$. The associated matrix is $\sum_{i=1}^{n} M_{[f_i, g_i]}$. Then

$$K = (\cap_{i=1}^{n} \ker f_i) \cap (\cap_{i=1}^{n} \ker g_i)$$

is infinite dimensional, and so for some nonzero $X \in Z^\vee$, $(\sum M_{[f_i, g_i]}) \cdot X = 0$. Since $M_0 \cdot X \neq 0$, it follows that $L^2 \subset_{\neq} L^{(2)}$.

In the general case, let $E \subset V$ be a countable subvector space, and let $\wedge T$ be the minimal model of $(\mathbb{Q} \oplus E, 0)$. Then its homotopy Lie algebra L_T is a retract of L

$$L_T \underset{\rho}{\overset{j}{\rightleftarrows}} L.$$

Let $\varphi_0 \in L_T^{(2)}$, not in L_T^2. Then $j(\varphi_0) \in L^{(2)}$ and not in L^2 because otherwise $\varphi_0 = \rho(\varphi)$ would belong to L_T^2.

$\square$

6.6   Free Products of Enriched Graded Lie Algebras

The category of enriched Lie algebras has free products. In fact the classical construction of the free product, $L \amalg L'$, of two graded Lie algebras extends naturally to enriched Lie algebras $(L, \{I_\alpha\})$ and $(L', \{I'_\beta\})$, with the enriched structure given by the surjections

$$\xi_{\alpha,\beta,n} : L \amalg L' \to L_\alpha \amalg L'_\beta / (L_\alpha \amalg L'_\beta)^n.$$

Its completion will be denoted by $L \,\widehat{\amalg}\, L'$, and almost by definition,

$$L \,\widehat{\amalg}\, L' \overset{\cong}{\longrightarrow} \overline{L} \,\widehat{\amalg}\, \overline{L'}.$$

Definition $L \,\widehat{\amalg}\, L'$ is the *free product* of L and L'.

Lemma 6.3

(i) With the notation above, any two morphisms $f : (L, \{I_\alpha\}) \to E$ and $g : (L', \{I'_\beta\}) \to E$ into an enriched Lie algebra extend uniquely to a morphism

$$L \mathbin{\widehat{\amalg}} L' \to E.$$

This identifies $L \mathbin{\widehat{\amalg}} L'$ as a coproduct in the category of complete Lie algebras and characterizes $L \mathbin{\widehat{\amalg}} L'$ up to natural isomorphism.

(ii) Suppose a profree Lie algebra $L = \overline{\mathbb{L}}_T$ and that $T = T(1) \oplus T(2)$ with $T(1)$ and $T(2)$ closed subspaces. Then $\overline{\mathbb{L}}_{T(1)}$ and $\overline{\mathbb{L}}_{(T(2)}$ are the closures of the Lie algebras generated by $T(1)$ and $T(2)$, and

$$\overline{\mathbb{L}}_{T(1)} \mathbin{\widehat{\amalg}} \overline{\mathbb{L}}_{T(2)} \xrightarrow{\;\cong\;} \overline{\mathbb{L}}_T.$$

Proof

(i) Let $\{J_\gamma\}$ denote the enriched structure for E. Then for each γ there are indices $\alpha(\gamma), \beta(\gamma)$ such that f and g factor to yield morphisms

$$f_\gamma : L_{\alpha(\gamma)} \to E_\gamma \quad \text{and } g_\gamma : L'_{\beta(\gamma)} \to E_\gamma.$$

Moreover, since $E_\gamma^{n(\gamma)} = 0$, some $n(\gamma)$, $f_\gamma \amalg g_\gamma$ factors to yield a morphism

$$h_\gamma : (L_{\alpha(\gamma)} \amalg L'_{\beta(\gamma)})/(L_{\alpha(\gamma)} \amalg L'_{\beta(\gamma)})^n \longrightarrow E_\gamma.$$

Since E is complete, these extend to

$$h := \varprojlim_\gamma h_\gamma : L \mathbin{\widehat{\amalg}} L' \to E.$$

The uniqueness is immediate.

(ii) The inclusion $T(1), T(2) \to T$ extends to a morphism $\overline{\mathbb{L}}_{T(1)} \mathbin{\widehat{\amalg}} \overline{\mathbb{L}}_{T(2)} \to \overline{\mathbb{L}}_T$, with inverse given by the extension to $\overline{\mathbb{L}}_T$ of $T = T(1) \oplus T(2) \to \overline{\mathbb{L}}_{T(1)} \mathbin{\widehat{\amalg}} \overline{\mathbb{L}}_{T(2)}$. $\square$

6.7   Limits and Colimits of Enriched Lie Algebras

Proposition 6.6 *The category $Enrich$ of enriched Lie algebras is complete and cocomplete, i.e., each diagram of enriched Lie algebras has a direct and an inverse limit.*

Proof The construction of the coproduct of two enriched Lie algebras can be generalized to any family $\{L(i)\}_{i \in I}$ of enriched Lie algebras. In fact, denote by $A_{i,\sigma}$ the family of ideals defining the enriched structure in $L(i)$. For each finite subset $J \subset I$, we denote by $p_J : \amalg_{i \in I} L(i) \to \amalg_{i \in J} L(i)$ the projection that is the identity on the components $L(j)$, $j \in J$, and is defined by $p_J(L(i)) = 0$ for $i \notin J$. Then for $J \subset I$ finite, and each ideal $A_{i,\sigma}$, and each integer n, we denote by $K_{i,\sigma,n}$ the kernel of the projection

$$\amalg_{i \in I} L(i) \xrightarrow{\ \ p_J\ \ } \amalg_{i \in J} L(i) \xrightarrow{\hspace{3cm}} \amalg_{i \in J} L(i)/A_{i,\sigma}^n .$$

These ideals define an enriched structure on $\amalg_{i \in I} L(i)$ and its completion $\overline{\amalg_{i \in I} L(i)}$ is the coproduct of the $L(i)$ is the category of enriched Lie algebras.

Now let $(L(i), f_j)$ be a diagram in *Enrich*. Its colimit in the category of Lie algebras is the quotient of $\amalg_{i \in I} L(i)$ by an ideal J associated with the morphisms f_j. Denote by $\iota : \amalg_{i \in I} L(i) \to \overline{\amalg_{i \in I} L(i)}$ the natural injection and by $\overline{J}$ the closure of the ideal generated by $\iota(J)$ in $\overline{\amalg_{i \in I} L(i)}$. Then by the corollary to Lemma 2.2, the quotient $\overline{\amalg_{i \in I} L(i)}/\overline{J}$ inherits an enriched structure. This is clearly the colimit of the diagram $(L(i), f_j)$ in *Enrich*.

Now consider a diagram $(L(i), f_j)$ of enriched Lie algebras. Its limit in the category of Lie algebras is the sub-Lie algebra $S \subset \prod L(i)$ consisting of families $a_i \in L(i)$ such that for any morphism $f_j : L(i_0) \to L(i_1)$ we have $f_j(a_{i_0}) = a_{i_1}$. Since for any morphism $f : L(1) \to L(2)$ of enriched Lie algebras, f is a limit of morphisms $f_\alpha : L(1)_{\beta(\alpha)} \to L(2)_\alpha$, the sub-Lie algebra S is closed and is the limit of the diagram in the category of enriched Lie algebras.

$\square$

Part II

Topological Spaces, Sullivan Rationalization, and Homotopy Lie Algebras

Recall that we identify the categories of topological spaces and simplicial sets via the singular simplex and Milnor realization functors [64].

Sullivan Rational Homotopy Theory

In this section we recall the basic constructions of rational homotopy theory, in particular the construction of the minimal Sullivan model of a space and its main properties. We recall also the elements of the model category structure defined on the connected differential graded algebras and the role of the minimal models in that structure.

7.1 Model Category Theory

Definition ([23,73]) A *model category* is a category C with three distinguished classes of maps: the *weak equivalences*, the *fibrations*, and the *cofibrations*, each of which is closed under composition and contains the identity maps. A map that is both a fibration and a weak equivalence is called an *acyclic fibration*. A map that is both a cofibration and a weak equivalence is called an *acyclic cofibration*. They must satisfy the following five axioms.

MC1 C is complete and cocomplete.

MC2 If f and g are maps in C such that gf is defined and if any two of the three maps f, g, gf are weak equivalences, then so is the third.

MC3 If f is a retract of g and if g is a fibration, a cofibration, or a weak equivalence, then so is f.

MC4 Given a commutative diagram with solid arrows

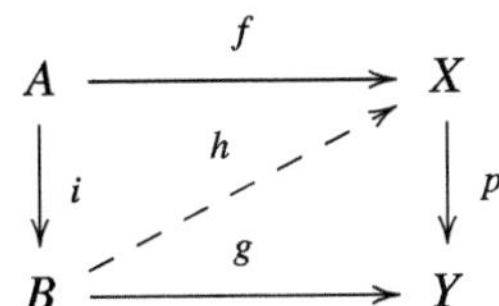

Y. Félix, S. Halperin, *Lie Models for Spaces*, Frontiers in Mathematics,
https://doi.org/10.1007/978-3-032-15357-9_7

there is a map (called a lift) $h : B \to X$ such that $hi = f$ and $ph = g$ either if i is a cofibration and p an acyclic fibration or if i is an acyclic cofibration and p is a fibration.

We say that i has the LLP (Left Lifting Property) with respect to p or that p has the RLP (Right Lifting Property) with respect to i.

MC5 Any map f can be factored in two ways: (1) $f = pi$, where i is a cofibration and p an acyclic fibration, and (2) $f = pi$, where i is an acyclic cofibration and p is a fibration.

For instance, set is a model category with [65]:

- The objects are pointed simplicial sets, and the morphisms are the maps preserving the base points.
- Weak equivalences are those maps whose realizations are weak homotopy equivalences.
- Cofibrations are inclusions.
- Fibrations are maps that have the right lifting property with respect to acyclic cofibrations.

Also, if $\mathcal{C}$ is a model category, then $\mathcal{C}^{op}$ is also a model category.

- Weak equivalences in $\mathcal{C}$ are weak equivalences in $\mathcal{C}^{op}$.
- Fibrations in $\mathcal{C}$ are cofibrations in $\mathcal{C}^{op}$.
- Cofibrations in $\mathcal{C}$ are fibrations in $\mathcal{C}^{op}$.

Finally, given model categories $\mathcal{C}$ and $\mathcal{D}$, a Quillen adjunction is an adjunction

$$F : \quad \mathcal{C} \underset{\longleftarrow}{\overset{\longrightarrow}{\rule{3em}{0pt}}} \mathcal{D} \quad : G$$

such that F preserves cofibrations and acyclic cofibrations or equivalently that G preserves fibrations and acyclic fibrations.

7.2   The A_{PL} and the Realization Functors

Sullivan's approach to rational homotopy theory is based on a two-way relation between topological spaces and Sullivan algebras [77], [36, §1.6], which we briefly review for the convenience of the reader. This relation is obtained from two contravariant adjoint functors, the $A_{PL}(-)$ functor and the geometric realization functor $\langle - \rangle$

$$A_{PL} : \text{sSet} \rightsquigarrow \text{Cdga} \quad \text{and} \quad \langle \ \rangle : \text{Cdga} \rightsquigarrow \text{sSet}.$$

These are constructed as follows from the simplicial cdga $A_{PL} = \{A_{PL}(\Delta^n)\}$:

$$A_{PL}(X) = \mathrm{sSet}(X, A_{PL}) \quad \text{and} \quad \langle A \rangle = \mathrm{Cdga}(A, A_{PL}).$$

Here the cdga structure of $A_{PL}(X)$ and the simplicial structure of $\langle A \rangle$ are inherited from the corresponding structures of A_{PL}. The adjointness of these functors follows from the following equalities for all cdga's A and simplicial sets X:

$$\mathrm{cdga}(A, A_{PL}(X)) = \mathrm{sSet}(X, \langle A \rangle). \tag{7.1}$$

Remark that if $(X, *)$ is a pointed simplicial set, then $A_{PL}(X)$ is naturally augmented by the projection $A_{PL}(X) \twoheadrightarrow A_{PL}(*)$.

The following class of morphisms of cdga's is important in the theory.

Definitions

- A *relative Sullivan algebra* consists of an inclusion of cdga's of the form $(A, d) \to (A \otimes \wedge V, d)$ such that V has a basis $v_{\alpha, \alpha \in A}$, where A is a well-ordered set such that $dv_\beta \in A \otimes \wedge V_\beta$, where V_β denotes the linear span generated by the v_α, $\alpha < \beta$.
- A Sullivan relative algebra $(\wedge V, d)$ extending $(A, d) = (\mathbb{Q}, 0)$ is called a *Sullivan algebra*.
- The Sullivan algebra $(\wedge V, d)$ is *minimal* if $dV \subset \wedge^{\geq 2} V$.

Moreover, each cdga admits a quasi-isomorphism $\varphi : (\wedge V, d) \xrightarrow{\simeq} (A, d)$ from a minimal Sullivan algebra; φ is the *minimal Sullivan model* of A and is unique up to homotopy class of isomorphism. (For simplicity, we denote $(\wedge W, d)$ simply by $\wedge W$ when we do not need to specify the differential.)

Theorem 7.1 ([11, 36]) *The category cdga is a model category, and the adjoint pair*

$$\langle \cdot \rangle : \quad \mathrm{Cdga} \; \rightleftarrows \; \mathrm{sSet}^{\mathrm{op}} \quad : A_{PL}$$

is a Quillen adjunction.

The model category of cdga is given by:

- Weak equivalences are quasi-isomorphisms.
- Fibrations are degreewise surjections.
- Cofibrations are retracts of relative Sullivan algebras.

Moreover cdga is a cofibrantly generated model category. The generating cofibrations are the morphisms

$$i_n : (\wedge x, 0) \to (\wedge(x, y), dy = x), \quad \deg x = n,$$

and the generating acyclic cofibrations are the morphisms

$$j_n : 0 \to (\wedge(x, y), dy = x), \qquad \deg x = n.$$

In summary, the two-way relation between topological spaces and Sullivan algebras is given by the following definition:

Definition

(i) A *(minimal) Sullivan model* of a connected space X is a quasi-isomorphism

$$\varphi : \wedge V \xrightarrow{\simeq} A_{PL}(X)$$

from a (minimal) Sullivan algebra.

(ii) A *Sullivan rationalization* of a connected space X is the adjoint map

$$\widetilde{\varphi} : X \to X_{\mathbb{Q}} = \langle \wedge V \rangle$$

of its minimal Sullivan model, φ.

We will usually use the notation $\widetilde{\varphi}$ for the adjoint of a minimal Sullivan model $\varphi :$ $\wedge V \to A_{PL}(X)$.

(iii) A *Sullivan representative of a continuous map* $f : X \to Y$ is a morphism $\varphi_f :$ $\wedge V_Y \to \wedge V_X$ between Sullivan models making commutative up to homotopy the diagram

$$
\begin{array}{ccc}
\wedge V_Y & \xrightarrow{\;\;\varphi_Y\;\;} & A_{PL}(Y) \\
\Big\downarrow{\varphi_f} & & \Big\downarrow{A_{PL}(f)} \\
\wedge V_X & \xrightarrow{\;\;\varphi_X\;\;} & A_{PL}(X).
\end{array}
$$

(iv) For any Sullivan model $\wedge V$, we denote $\iota_{\wedge V} : \wedge V \to A_{PL}\langle \wedge V \rangle$ the map adjoint to $id_{\langle \wedge V \rangle} : \langle \wedge V \rangle \to \langle \wedge V \rangle$.

Definition If $(\wedge V, d)$ is a Sullivan algebra, then two morphisms $\varphi_0, \varphi_1 : \wedge V \to \wedge W$ are *homotopic* if there is a morphism of cdga $\Phi : \wedge V \to \wedge W \otimes \wedge(t, dt)$ such that $\Phi(Z) \subset \wedge^{\geq 1} W \otimes \wedge(t, dt)$, and for $i = 0, 1$, $\varphi_i = (1 \otimes \varepsilon_i) \circ \Phi$.

Remark This definition is called *based homotopy* in [36]. This is the good notion corresponding to based homotopy of pointed simplicial sets.
The requirement on the image of $\Phi(Z)$ is important as shown by the following example:

Suppose $\wedge V = \wedge W = \wedge(v, v_1, v_2)$, $dv = v_1 v_2$, $dv_1 = dv_2 = 0$, v, v_1, v_2 in degree 1. Then the map $\varphi_1 : \wedge V \to \wedge V$ defined by $\varphi_1(v_1) = v_1$, $\varphi_1(v_2) = v_2$, and $\varphi_1(v) = v + v_1$ is not homotopic to $\varphi_0 = id_{\wedge V}$. There is anyway a morphism $\Phi : \wedge V \to \wedge W \otimes \wedge(t, dt)$ with $\varphi_i = (1 \otimes \varepsilon_i) \circ \Phi$ for $i = 0, 1$

$$\Phi(v_1) = v_1, \quad \Phi(v_2) = v_2 - dt, \quad \Phi(v) = v + tv_1.$$

Remark The uniqueness (up to homotopy class of quasi-isomorphisms) of a Sullivan model implies the uniqueness (up to homotopy class of homotopy equivalences) of a Sullivan rationalization.

A general connection between $\iota_{\wedge V}$ and $\widetilde{\varphi}_X$ is given in the following lemma:

Lemma 7.1 *Let* $\iota_{\wedge V} : \wedge V \to A_{PL}\langle \wedge V \rangle$ *adjoint to* $id_{\langle \wedge V \rangle}$. *Then* $\varphi \sim A_{PL}(\widetilde{\varphi}) \circ \iota_{\wedge V}$.

$$\varphi = A_{PL}(\widetilde{\varphi}) \circ \iota_{\wedge V} : \quad \wedge V \xrightarrow{\ \iota_{\wedge V}\ } A_{PL}\langle \wedge V \rangle \xrightarrow{\ A_{PL}(\widetilde{\varphi})\ } A_{PL}(X) \qquad (7.2)$$

Proof By adjunction from the commutative diagram

$$
\begin{array}{ccc}
\langle \wedge V \rangle & \xleftarrow{\ \widetilde{\varphi}\ } & X \\
\Big\| = & & \Big\downarrow \widetilde{\varphi} \\
\langle \wedge V \rangle & \xrightarrow{\ = \ } & \langle \wedge V \rangle
\end{array}
$$

we deduce the commutative diagram

$$
\begin{array}{ccc}
\wedge V & \xrightarrow{\ \varphi\ } & A_{PL}(X) \\
= \Big\uparrow & & \Big\uparrow A_{PL}(\widetilde{\varphi}) \\
\wedge V & \xrightarrow{\ \iota_{\wedge V}\ } & A_{PL}\langle \wedge V \rangle
\end{array}
$$

$\square$

We also note that immediately from the definitions we obtain the following proposition:

Proposition 7.1 *Suppose* $\wedge V$ *and* $\wedge W$ *are Sullivan algebras. Then:*

(i)   The natural map

$$\langle \wedge V \otimes \wedge W \rangle \to \langle \wedge V \rangle \times \langle \wedge W \rangle$$

is a homotopy equivalence.

(ii) The diagram

$$
\begin{array}{ccccc}
\wedge V \otimes \wedge W & \xrightarrow{\iota_{\wedge V} \otimes \iota_{\wedge W}} & A_{PL}\langle\wedge V\rangle \otimes A_{PL}\langle\wedge W\rangle & \longleftarrow & A_{PL}(\langle\wedge V\rangle \times \langle\wedge W\rangle) \\
\downarrow & & \downarrow & & \downarrow \\
\wedge V \times_{\mathbb{Q}} \wedge W & \longrightarrow & A_{PL}\langle\wedge V\rangle \times_{\mathbb{Q}} A_{PL}\langle\wedge W\rangle & \xleftarrow{\simeq} & A_{PL}(\langle\wedge V\rangle \vee \langle\wedge W\rangle)
\end{array}
$$

commutes. If either $H(\langle\wedge V\rangle)$ or $H(\langle\wedge W\rangle)$ is a graded vector space of finite type, then $A_{PL}\langle\wedge V\rangle \otimes A_{PL}\langle\wedge W\rangle \longleftarrow A_{PL}(\langle\wedge V\rangle \times \langle\wedge W\rangle)$ is also a quasi-isomorphism.

7.3 The Homotopy Groups $\pi_*(X_{\mathbb{Q}})$

Proposition 7.2 *Let $(\wedge V, d)$ be a minimal Sullivan algebra; then for each $n \geq 1$, there is a natural isomorphism of abelian groups*

$$
\tau_V : \pi_n\langle\wedge V\rangle \xrightarrow{\cong} (V^n)^{\vee}.
$$

Proof It is straightforward from the definitions that (7.1) induces a natural bijection

$$
[\wedge V, A_{PL}(S^n)] \cong [S^n, \langle\wedge V\rangle], \qquad [\varphi] \mapsto [\widetilde{\varphi}], \tag{7.3}
$$

of based homotopy classes of morphisms.

If $\wedge W(n)$ is the minimal model of the n-sphere S^n, then we have natural cdga quasi-isomorphisms

$$
A_{PL}(S^n) \xleftarrow{\simeq} \wedge W(n) \xrightarrow{\simeq} H(S^n) = \mathbb{Q} \oplus \mathbb{Q}a_n,
$$

a_n denoting the orientation class of S^n. Therefore (12) yields

$$
\tau_V : \pi_n\langle\wedge V\rangle = [S^n, \langle\wedge V\rangle] = [\wedge V, H(S^n)] = (V^n)^{\vee}. \tag{7.4}
$$

Moreover, a morphism $\psi : \wedge V \to \wedge Z$ of minimal Sullivan algebras induces, via division by $\wedge^{\geq 2}$, a linear map $Q(\psi) : V \to Z$. It is immediate that

$$
\tau_V \circ \pi_*\langle\psi\rangle = Q(\psi)^{\vee} \circ \tau_Z. \tag{7.5}
$$

$$
\begin{array}{ccc}
\pi_*\langle\wedge Z\rangle & \xrightarrow{\pi_*\langle\psi\rangle} & \pi_*\langle\wedge V\rangle \\
\downarrow{\scriptstyle \tau_Z} & & \downarrow{\scriptstyle \tau_V} \\
Z^{\vee} & \xrightarrow{Q(\psi)} & V^{\vee}
\end{array}
$$

Thus τ_V naturally endows $\pi_*\langle \wedge V\rangle$ with the structure of a graded rational vector space. Now, by [36, Theorem 1.4] for $n \geq 2$, τ_V is an isomorphism of abelian groups. $\qquad\square$

Proposition 7.3 *Let $f : X \to Y$ be a continuous map between connected spaces with Sullivan representative $\varphi : \wedge V_Y \to \wedge V_X$. Then the following are equivalent:*

(i) $H(f)$ is an isomorphism (f is a rational homotopy equivalence).
(ii) $\langle \varphi \rangle : \langle \wedge V_Y \rangle \to \langle \wedge V_X \rangle$ is a homotopy equivalence.

Proof When f is a rational homology equivalence, then X and Y have the same minimal Sullivan model, and so $\langle \varphi \rangle$ is a homotopy equivalence.

When $\langle \varphi \rangle$ is a homotopy equivalence, then the induced map $V_X^\vee \to V_Y^\vee$ is an isomorphism, which implies that $H_*(f)$ is an isomorphism, and so f is a rational homology equivalence. $\qquad\square$

The Homotopy Lie Algebra L_V of a Minimal Sullivan Algebra

8

To each minimal Sullivan algebra $\wedge V$ is associated with an enriched Lie algebra L_V, isomorphic to the homotopy Lie algebra of the geometric realization $\langle \wedge V \rangle$ of $\wedge V$. When $\wedge V$ is the minimal model of a connected space X, then $\langle \wedge V \rangle$ is the rationalization $X_{\mathbb{Q}}$ of X, and L_V is denoted by L_X. We present the first properties of L_V and describe L_X in the case of a wedge of spaces.

8.1 The Homotopy Lie Algebra L_V

Associated with a minimal Sullivan algebra, $(\wedge V, d)$ is the quadratic Sullivan algebra $(\wedge V, d_1)$ defined as follows: $d_1 v$ *is the component in* $\wedge^2 V$ *of* dv. This in turn determines (Proposition 4.1) the enriched Lie algebra $L_V = (L_V)_{\geq 0}$ given by

$$s L_V = V^{\vee} \quad \text{(as a graded vector space)}$$

and

$$< v, s[x, y] >= (-1)^{1+deg\, y} < d_1 v, sx, sy > .$$

Recall (Chap. 4) that L_V was introduced as the homotopy Lie algebra of $(\wedge V, d_1)$. Now we extend this to $(\wedge V, d)$ in the following definition:

Definition L_V is the *homotopy Lie algebra of* $(\wedge V, d)$.
If $(\wedge V, d)$ is the minimal Sullivan model of a connected space, X, then L_V is the *homotopy Lie algebra of* X and is often denoted L_X

$$L_X \cong \pi_*(X_{\mathbb{Q}}).$$

© The Author(s), under exclusive license to Springer Nature Switzerland AG 2026
Y. Félix, S. Halperin, *Lie Models for Spaces*, Frontiers in Mathematics,
https://doi.org/10.1007/978-3-032-15357-9_8

Lemma 8.1 *L_V is a natural enriched graded Lie algebra.*

Proof Write $(\wedge V, d) = \varinjlim_\sigma (\wedge V_\sigma, d)$, where $\dim V_\sigma < \infty$ for all σ. Then $L_V = \varprojlim_\sigma L_{V_\sigma}$ is an enriched graded Lie algebra. $\qquad\square$

Finally, a minimal Sullivan algebra $(\wedge V, d)$ has a natural filtration given by the ideals $\wedge^{\geq k} V$. The first term in the corresponding spectral sequence is a quadratic Sullivan algebra $(\wedge V, d_1)$, and if $\varphi : (\wedge V, d) \to (\wedge W, d)$ is a morphism of minimal Sullivan algebras, it induces a morphism $\varphi_1 : (\wedge V, d_1) \to (\wedge W, d_1)$ in which $\varphi_1 v$ is the component of φv in W.

Proposition 8.1 *Suppose $\varphi, \psi : (\wedge V, d) \to (\wedge W, d)$ are homotopic morphisms between minimal Sullivan algebras. Then:*

  (i) The induced morphisms $\varphi_1, \psi_1 : (\wedge V, d_1) \to (\wedge W, d_1)$ coincide.
  (ii) The induced morphisms $L_V \leftarrow L_W$ coincide.
 (iii) $\varphi|_{V^1} = \psi|_{V^1}$.

Proof

  (i) Suppose $\Phi : (\wedge V, d) \to (\wedge W, d) \otimes \wedge(t, dt)$ is a homotopy from φ to ψ. Then by definition $\Phi(V) \subset \wedge^{\geq 1} W \otimes \wedge(t, dt)$. Filtering by the ideals $\wedge^{\geq j} V$ and $\wedge^{\geq j} W$ provides a homotopy

$$\Gamma : (\wedge V, d_1) \to (\wedge W, d_1) \otimes \wedge(t, dt)$$

from φ_1 to ψ_1. Moreover, division by $\wedge^{\geq 2} V$ and $\wedge^{\geq 2} W$ converts Γ into a chain homotopy between

$$\varphi_1|_V, \psi_1|_V : (V, 0) \to (W, 0).$$

  It follows that $\varphi_1|_V = \psi_1|_V$ and therefore that $\varphi_1 = \psi_1$.
  (ii) This is immediate from (i).
 (iii) This follows because $\varphi|_{V^1} = \varphi_1|_{V^1} = \psi_1|_{V^1} = \psi|_{V^1}$. $\qquad\square$

Now, for any minimal Sullivan algebra $\wedge V$, the identifications (7.4) extend to natural bijections

$$\tau_V : \pi_n\langle \wedge V \rangle \xrightarrow{\;\cong\;} (V^n)^\vee = s(L_V)_{n-1}. \tag{8.1}$$

This converts the Lie brackets in L_V to the Whitehead products in $\pi_*\langle \wedge V \rangle$.

Proposition 8.2 *Suppose $(\wedge V, d)$ is a minimal Sullivan algebra.*

(i) *For $n \geq 2$, the bijection τ_V is a linear isomorphism. In particular, $\pi_n\langle\wedge V\rangle$ is the rational vector space $s(L_V)_{n-1}$.*

(ii) *For $p, q \geq 2$, (8.1) is an isomorphism of graded Lie algebras identifying the Lie bracket $(L_V)_p \times (L_V)_q \xrightarrow{[\,,\,]} (L_V)_{p+q}$ with the Whitehead product $\pi_{p+1}\langle\wedge V\rangle \times \pi_{q+1}\langle\wedge V\rangle \to \pi_{p+q+1}\langle\wedge V\rangle$.*

(iii) *The composite*

$$\pi_1\langle\wedge V\rangle = s(L_V)_0 \xrightarrow[\cong]{\exp} G_L$$

is an isomorphism of groups, where G_L is the fundamental group of L_V defined in Sect. 3.1.

(iv) *Finally τ_V identifies the right adjoint representation of $\pi_1\langle\wedge V\rangle$ in $\pi_n\langle\wedge V\rangle$ as*

$$\beta \bullet \exp\alpha = e^{-ad\,\alpha}(\beta), \qquad \alpha \in (L_V)_0, \beta \in L_V.$$

Proof Recall that $(\wedge V, d) = \varinjlim_\alpha (\wedge V_\alpha, d)$ in which the V_α are finite dimensional subspaces of V. It follows that $\pi_*\langle\wedge V\rangle = \varprojlim_\alpha \langle\wedge V_\alpha\rangle$. Now Proposition 8.2 follows, respectively, from [36, Theorem 1.4], [33, Proposition 13.16], [36, Theorem 2.4], and [36, Theorem 2.5]. $\square$

Moreover, a continuous map $f : Y \to X$ has a unique homotopy class of Sullivan representatives $\varphi : \wedge V_Y \leftarrow \wedge V_X$ between the respective minimal Sullivan models, inducing $f_\mathbb{Q} : Y_\mathbb{Q} \to X_\mathbb{Q}$ and a Lie algebra morphism $L_f : L_Y \to L_X$. It follows from (8.1) that

$$\pi_*(f_\mathbb{Q}) = \pi_*\langle\varphi\rangle = sL_f. \tag{8.2}$$

Example Denote by $\wedge V$ the minimal Sullivan model of $S^1 \vee S^3$. Then $V^1 = \mathbb{Q}v$, while V^3 has a basis $w_0, w_1, \ldots, w_n, \ldots$ satisfying $dw_0 = 0$ and $dw_{n+1} = vw_n$. Define $x \in (L_V)_0$ by $<v, sx> = 1$, and identify $\pi_3\langle\wedge V\rangle = (V^3)^\vee$ as the space of infinite series

$$g(t) = \sum_{n\geq 0} c_n t^n$$

with $<w_n, sg(t)> = c_n$. Then the action of $\pi_1\langle\wedge V\rangle$ in $\pi_3\langle\wedge V\rangle$ is given by

$$g(t) \bullet \exp x = e^t \cdot g(t).$$

$\square$

The Hurewicz Homomorphism

The Hurewicz homomorphism $hur : \pi_*(X) \to H_*(X; \mathbb{Z})$ provides another means of connecting minimal Sullivan models to topology. In fact, let

$$\psi : \wedge V \to A_{PL}(Y)$$

be any morphism to the PL forms on a connected space. Then it determines the diagram

$$
\begin{array}{ccc}
\pi_*(Y) & \xrightarrow{\;\;\pi_*(\widetilde{\psi})\;\;} & V^\vee \\
\Big\downarrow{\scriptstyle hur} & & \Big\downarrow{\scriptstyle H(\xi)^\vee} \\
H_*(Y; \mathbb{Z}) & \xrightarrow{\;\;\omega_H\;\;} & H^{\geq 1}(\wedge V)^\vee,
\end{array}
\tag{8.3}
$$

defined as follows:

- $\xi : \wedge^{\geq 1} V \to V$ is division by $\wedge^{\geq 2} V$.
- ω_H is the composite

$$H_*(Y; \mathbb{Z}) \to H_*(Y; \mathbb{Q}) \to H(Y)^\vee \xrightarrow{\;H(\psi)^\vee\;} H(\wedge V)^\vee.$$

Now [36, Proposition 1.19] can be restated as follows:

Proposition 8.3 *Diagram (8.3) is commutative.*

Remark In the specific example in which $\psi = \iota_{\wedge V} : \wedge V \to A_{PL}\langle \wedge V \rangle$, diagram (8.3) reduces to

$$
\begin{array}{ccc}
\pi_*\langle \wedge V \rangle & \xrightarrow{\;\;\cong\;\;} & V^\vee \\
\Big\downarrow{\scriptstyle hur} & & \Big\downarrow{\scriptstyle H(\xi)^\vee} \\
H_*\langle \wedge V \rangle & \xrightarrow{\;H(\iota_{\wedge V})^\vee\;} & H^{\geq 1}(\wedge V)^\vee.
\end{array}
$$

Lemma 8.2

(i) Let L be the homotopy Lie algebra of a simply connected CW complex, Y of finite type. If $L_2 \not\subset [L_1, L_1]$, then the Hurewicz map $\pi_3(Y) \to H_3(Y)$ is nonzero.

(ii) Let $X = \langle \wedge V \rangle$ be the spatial realization of the Sullivan quadratic model of the profree Lie algebra $L = \overline{\mathbb{L}}_T$, where $T = T_1$ is an infinite dimensional vector space. Then $H_3(X) \neq 0$.

Proof

(i) Denote by $(\wedge V, d)$ the Sullivan minimal model of Y. Then V is a vector space of finite type. Therefore the bracket $[\,,\,] : \wedge^2 L_1 \to L_2$ is dual to the differential $d : V^3 \to \wedge^2 V^2$. Since $L_2 \neq [L_1, L_1]$, $d_{|V^3}$ is not injective, and $H^3(\wedge V) \to H^3(\wedge V / \wedge^{\geq 2} V)$ is nonzero. But the map $V^3 \cap d \to V^3$ is the dual of the Hurewicz map $\pi_3(Y) \to H_3(Y; \mathbb{Q})$. Thus hur $(\omega) \neq 0$.

(ii) Recall (Lemma 4.1) that $V = V^{\geq 2}$. Because (12), $\pi_*\langle \wedge V \rangle = sL$. Now let $\omega \in \pi_3(X)$ be a homotopy class corresponding by suspension to an element in L_2, but not in $[L_1, L_1]$. Then X contains a simply connected finite CW complex Y such that ω is in the image of $\pi_3(Y)$. We denote this element by $\omega_Y \in \pi_3(Y)$. It follows that ω_Y is not decomposable, and by Lemma 8.2(i), the image a_Y of ω_Y in $H_3(Y)$ is nonzero.

For any finite sub-CW complex Z of X containing Y, the element ω_Z induces also a nonzero element $a_Z \in H_3(Z)$ by the Hurewicz map. By naturalness of the Hurewicz map, the map $H_3(Y) \to H_3(Z)$ maps a_Y to a_Z. Now X is the union of the finite sub-CW complexes Z_α containing Y. It follows that $H_3(X) = \varinjlim_\alpha H_3(Z_\alpha)$, and therefore the family a_{Z_α} induces a nonzero element in $H_3(X)$. $\square$

8.2   A Countable Property of L_V

Proposition 8.4 *The following conditions on a minimal Sullivan algebra $(\wedge V, d)$ are equivalent:*

(i) *$L_V = \varprojlim_{n \in \mathbb{N}} L(n)$ in which each $L(n)$ is a finite dimensional nilpotent Lie algebra.*
(ii) *$\dim V$ is at most countable.*
(iii) *$\dim H(\wedge V, d)$ is at most countable.*

Proof (i) $\Rightarrow$ (ii). Let $\wedge V_n$ be the quadratic Sullivan model of $L(n)$. Then $(\wedge V, d_1) = \varinjlim_n (\wedge V_n, d_1)$ is at most countable. This proves (ii) and (ii) $\Rightarrow$ (iii) is obvious.

(iii) $\Rightarrow$ (i). Define subspaces $W_n \subset V$ by $W_0 = V \cap \ker d$ and $W_{n+1} = V \cap d^{-1}(\wedge W_n)$. It follows from (iii) that each $\dim W_n$ is at most countable. Since $(\wedge V, d)$ is a Sullivan algebra, it follows that $V = \cup_n W_n$, and so $\dim V$ is at most countable. Then write $V = \cup_\alpha V_\alpha$ where $\dim V_\alpha < \infty$ and $\wedge V_\alpha$ is preserved by d. Since $\dim V$ is at most countable, this implies that $V = \varinjlim_n V_n$, where $(\wedge V_n, d)$ is a sub-Sullivan algebra and $\dim V_n < \infty$; thus $L_V = \varprojlim_n L(n)$, $L(n)$ denoting the homotopy Lie algebra of $(\wedge V_n, d)$. $\square$

Proposition 8.5 *Let $(\wedge V, d)$ be a minimal Sullivan algebra, and suppose that $V = V^{\geq 2}$ is countable. Then we have an isomorphism $H_*(\Omega\langle \wedge V\rangle) \cong \overline{UL}$.*

Proof Since $H(\wedge V, d)$ is at most countable, $(\wedge V, d)$ is the union of an increasing sequence of minimal Sullivan algebras $(\wedge V(n), d)$ with $\dim V(n) < \infty$. It follows that $\langle \wedge V\rangle$ is the homotopy inverse limit of the tower $Z(n) \to Z(n-1)$, where $Z(n) = \langle \wedge V(n)\rangle$. Therefore $\Omega\langle \wedge V\rangle = \varprojlim_n \Omega Z(n)$. Denote by $L(n)$ the homotopy Lie algebra of $Z(n)$.

Now since $V = V^{\geq 2}$, $H_*(\Omega Z(n))$ is a nilpotent algebra of the form $UL(n)$ where $\dim L(n) < \infty$. There is then a natural morphism of algebras

$$\varphi : H_*(\Omega Z) \to \varprojlim_n H_*(\Omega Z(n)).$$

Forgetting the multiplication, $\Omega Z(n)$ is homotopy equivalent to a product of Eilenberg-MacLane spaces. It follows that we can endow $H_*(\Omega Z(n))$ the structure of an abelian Hopf algebra. Now since the morphisms $H_*(\Omega Z(n)) \to H_*\Omega Z(n-1))$ are surjective, it follows from [38, Theorem B] that with those new structures the induced map

$$\varphi : H_*(\Omega Z) \xrightarrow{\cong} \varprojlim_n H_*(\Omega Z(n))$$

is an isomorphism of Hopf algebras and in particular an isomorphism of coalgebras.
On the other hand,

$$\overline{UL_V} = \varprojlim_n \overline{UL(n)} = \varprojlim_n UL(n) = \varprojlim_n H_*\Omega Z(n).$$

This implies the result.

$\square$

8.3   Spaces with Profree Homotopy Lie Algebra

Profree Lie algebras are characterized in Chap. 6 in terms of their quadratic Sullivan models. This can, however, be extended to general minimal Sullivan algebras (Proposition 8.6 below), and they arise, with their sub-Lie algebras, in a natural way as homotopy Lie algebras of wedges of spheres.

Proposition 8.6 *Denote by L and by $(\wedge V, d_1)$ the homotopy Lie algebra and the associated quadratic Sullivan algebra of a minimal Sullivan algebra $(\wedge V, d)$. The following conditions are then equivalent:*

(i) *There is a quasi-isomorphism* $(\wedge V, d) \xrightarrow{\simeq} \mathbb{Q} \oplus S$ *with zero differential in S and* $S \cdot S = 0$.

(ii) *The generating space V can be chosen so that $d : V \to \wedge^2 V$ and*

$$\mathbb{Q} \oplus (V \cap \ker d) \xrightarrow{\cong} H(\wedge V, d).$$

(iii) $L = \overline{\mathbb{L}}_T$ *is profree, with $sT = (H^{\geq 1}(\wedge V))^{\vee}$ defining the enriched structure.*

If they hold, then $(\wedge V, d) \cong (\wedge V, d_1)$, and V can be chosen in order that $V \cap \ker d = V \cap \ker d_1$.

Proof (i) $\Leftrightarrow$ (ii). First, suppose there is a quasi-isomorphism

$$(\wedge V, d) \xrightarrow{\simeq} (\mathbb{Q} \oplus S, 0).$$

A successive adjoining of variables to S constructs a quadratic Sullivan algebra $(\wedge W, d_1)$ such that $H^{[1]}(\wedge W, d_1) = S$ and $H^{[2]}(\wedge W) = 0$. (Here $H^{[k]}(\wedge W, d_1)$ denotes the homology classes represented by cycles in $\wedge^k W$.) Now Lemma 6.2 asserts that $H^{[k]}(\wedge W) = 0$, $k \geq 2$. Thus division by $\wedge^{\geq 2} W$ and by a direct summand of S in W defines a quasi-isomorphism, $(\wedge W, d_1) \xrightarrow{\simeq} \mathbb{Q} \oplus S$. Since minimal models are unique, it follows that there is an isomorphism

$$(\wedge W, d_1) \cong (\wedge V, d).$$

Now choose V, so this isomorphism takes $W \xrightarrow{\cong} V$, and by construction $S = V \cap \ker d$ and $H(\wedge V) \cong \mathbb{Q} \oplus (V \cap \ker d)$.

Now, if (ii) holds, then with the given choice of V, division by $\wedge^{\geq 2} V$ and by a direct summand of $V \cap \ker d$ in V defines a quasi-isomorphism

$$(\wedge V, d) \xrightarrow{\simeq} \mathbb{Q} \oplus (V \cap \ker d).$$

This gives (i).

The equivalence (i) $\Longrightarrow$ (iii) is Theorem 6.1.

$\square$

Now recall that a space X is *formal* if its minimal Sullivan model is also a minimal Sullivan model of $(H(X), 0)$. Then Proposition 8.6 has the following consequences:

Corollary 8.1 *If $H^{\geq 1}(X) \neq 0$, then the homotopy Lie algebra L_X is profree if and only if X is formal and $H^{\geq 1}(X) \cdot H^{\geq 1}(X) = 0$.*

Recall also that *the Lusternik-Schnirelmann category* of a space X, cat X, is the least integer n (or ∞) such that X can be covered by $n + 1$ open sets, each contractible in X.

Corollary 8.2 *If cat $X = 1$, then the homotopy Lie algebra, L_X, is profree.*

Proof It follows from [36, Formula (9.1) and Theorem 9.1] that if cat $X = 1$, then its minimal Sullivan model satisfies condition (i) of Proposition 8.6. $\square$

Example of Profree Lie Algebras

Let L be an enriched graded Lie algebra, and let $(\wedge V, d_1)$ be its associated quadratic Sullivan algebra, $d_1 : V \to \wedge^2 V$. Fix an integer $m \geq 1$. By [36, Proposition 9.4, Step Three] there are a graded vector space Z_m and a quasi-isomorphism

$$\varphi : (\wedge V \otimes (\mathbb{Q} \oplus Z_m), d) \xrightarrow{\simeq} (\wedge V / \wedge^{>m} V, \overline{d_1}),$$

where $d : Z_m \to \wedge^{m+1} V \oplus (V \otimes Z_m)$. Denote by $(\wedge V \otimes \wedge T_m)$ the Sullivan minimal model of $(\wedge V / \wedge^{>m} V, \overline{d_1})$.

It follows that $(\wedge T_m, \overline{d}) := \mathbb{Q} \otimes_{\wedge V} (\wedge V \otimes \wedge T_m)$ is quasi-isomorphic to $(\mathbb{Q} \oplus Z_m, 0)$. Therefore $\pi_*\langle \wedge T_m, \overline{d} \rangle$ is a profree Lie algebra F_m, and we have a short exact sequence of enriched Lie algebras

$$0 \to F_m \to \pi_*\langle \wedge V \otimes \wedge T_m \rangle \to L \to 0.$$

8.4   The Minimal Sullivan Model of $X \vee Y$

If X and Y are connected spaces, then the surjections $H(X \vee Y) \to H(X), H(Y)$ define an isomorphism $H(X \vee Y) \xrightarrow{\cong} H(X) \times_{\mathbb{Q}} H(Y)$. It follows that minimal Sullivan models $\wedge W \xrightarrow{\simeq} A_{PL}(X)$ and $\wedge Q \xrightarrow{\simeq} A_{PL}(Y)$ extend to a quasi-isomorphism

$$\wedge W \times_{\mathbb{Q}} \wedge Q \xrightarrow{\simeq} A_{PL}(X \vee Y).$$

This identifies a Sullivan model $\wedge T \xrightarrow{\simeq} \wedge W \times_{\mathbb{Q}} \wedge Q$ as a Sullivan model for $X \vee Y$.

On the other hand, for any two Sullivan algebras $\wedge W$ and $\wedge Q$, the surjection $\wedge W \otimes \wedge Q \to \wedge W \times_{\mathbb{Q}} \wedge Q$ extends to a quasi-isomorphism

$$\varphi : \wedge T := \wedge W \otimes \wedge Q \otimes \wedge R \xrightarrow{\simeq} \wedge W \times_{\mathbb{Q}} \wedge Q$$

from a minimal Sullivan extension. Since $\wedge W \otimes \wedge Q \to \wedge W \times_{\mathbb{Q}} \wedge Q$ is surjective, we may choose $R \subset T$ so that $\varphi(R) = 0$. It follows that $\wedge T$ is a minimal Sullivan algebra and that the quotient $(\wedge R, \overline{d})$ is a minimal Sullivan algebra.

Finally denote by $\wedge W \otimes \wedge U_W$ and $\wedge Q \otimes \wedge U_Q$ acyclic closures for $\wedge W$ and $\wedge Q$.

Proposition 8.7 *With the notation and hypotheses above:*

(i) The homotopy Lie algebra L_R is profree, $L_R = \mathbb{L}_S$ with $sS = (R \cap \ker \overline{d})^{\vee}$, and

$$R \cap \ker \overline{d} \xrightarrow{\;\cong\;} H^{\geq 1}(\wedge R, \overline{d}) \cong d(\wedge U_W) \otimes \wedge^{\geq 1} U_Q \cong s\left(\wedge^{\geq 1} U_W \otimes \wedge^{\geq 1} U_Q\right).$$

(ii) If $\wedge W$ and $\wedge Q$ are quadratic Sullivan algebras, then $\wedge T$ is quadratic.
(iii) In general, the quadratic Sullivan algebra $(\wedge T, d_1)$ for $\wedge T$ is the Sullivan model of $(\wedge W, d_1) \times_{\mathbb{Q}} (\wedge Q, d_1)$.
(iv) The inclusion $\wedge W \otimes \wedge Q \to \wedge T$ induces a short exact sequence

$$L_R \to L_T \to L_W \times L_Q$$

of enriched Lie algebras.

Remark Recall that the homotopy fiber F of the injection $X \vee Y \to X \times Y$ is the join of ΩX and ΩY and is thus a suspension. Proposition 8.7 is quite similar: The homotopy fiber F' of the inclusion $(X \vee Y)_{\mathbb{Q}} \to (X \times Y)_{\mathbb{Q}}$ has the rational homotopy type of a wedge of spheres.

Proof

(i) The combination of the acyclic closures and the morphism $\varphi : \wedge T \xrightarrow{\;\cong\;} \wedge W \times_{\mathbb{Q}} \wedge Q$ yields the quasi-isomorphisms

$$\wedge R \xleftarrow{\;\cong\;} \wedge T \otimes_{\wedge W \otimes \wedge Q} (\wedge W \otimes \wedge U_W \otimes \wedge Q \otimes \wedge U_Q) \xrightarrow{\;\cong\;} A := (\wedge W \times_{\mathbb{Q}} \wedge Q)$$
$$\otimes \wedge U_W \otimes \wedge U_Q.$$

Now divide A by the ideal generated by W to obtain the short exact sequence

$$0 \to \wedge^{\geq 1} W \otimes \wedge U_W \otimes \wedge U_Q \to A \to \wedge Q \otimes \wedge U_W \otimes \wedge U_Q \to 0.$$

Next, decompose the differential in $\wedge W \otimes \wedge U_W$ in the form $d = d_1 + d'$ with $d_1(W) \subset \wedge^2 W$, $d_1(U_W) \subset W \otimes \wedge U_W$, $d'(W) \subset \wedge^{\geq 3} W$, and $d'(U_W) \subset \wedge^{\geq 2} W \otimes \wedge U_W$. Then d_1 is a differential, and $(\wedge W \otimes \wedge U_W, d_1)$ is the acyclic closure of $(\wedge W, d_1)$. Choose a direct summand, S, of $d_1(\wedge^{\geq 1} U_W)$ in $W \otimes \wedge U_W$. Then $I = (\wedge^{\geq 2} W \otimes \wedge U_W) \oplus S$ is acyclic for the differential d_1 and therefore also for the differential d. Thus $J = I \otimes \wedge U_Q$ is an acyclic ideal in A and $A \xrightarrow{\;\cong\;} A/J$.

On the other hand, consider the short exact sequence

$$0 \to d(\wedge U_W) \otimes \wedge U_Q \to A/J \to \wedge Q \otimes \wedge U_W \otimes \wedge U_Q \to 0$$

in which $d(\wedge U_W) \otimes \wedge U_Q$ is an ideal with trivial multiplication and trivial differential. Now $H(\wedge Q \otimes \wedge U_W \otimes \wedge U_Q) \cong \mathbb{Q} \oplus \wedge^{\geq 1} U_W$, and the connecting map sends $\wedge^{\geq 1} U_W$ onto $d(\wedge U_W) \otimes \mathbb{Q}$. It follows from the long homology sequence that

$$\mathbb{Q} \oplus \left(d(\wedge U_W) \otimes \wedge^{\geq 1} U_Q \right) \xrightarrow{\simeq} A/J.$$

Since $\wedge R \simeq A \simeq A/J$, by Proposition 8.6, this completes the proof of (i).

Finally, since $d : \wedge^{\geq 1} U_W \to \wedge W \otimes \wedge U_W$ is injective, the morphism $d : \wedge^{\geq 1} U_W \to d(\wedge U_W)$ is an isomorphism.

(ii) Assign $\wedge W$ and $\wedge Q$ wedge degree as a second degree and assign U_W and U_Q second degree 0. Then $(\wedge W \otimes \wedge U_W, d_1)$ and $(\wedge Q \otimes \wedge U_Q, d_1)$ are the respective acyclic closures of $(\wedge W, d_1)$ and $(\wedge Q, d_1)$, and d_1 increases the second degree by 1. Now φ and T may be constructed so that R is equipped with a second gradation for which d increases the second degree by one and φ is bihomogeneous of degree zero, and $(\wedge T, d)$ is quadratic.

(iii) Starting from the quadratic Sullivan model $(\wedge T, d_1)$, we construct a differential d in $\wedge T$ that is an extension of the differential in $\wedge U \otimes \wedge W$ and that satisfies $d = d_1 + d_2 + \dots$, where d_r increases the new gradation by r. We construct $d_2, d_3, \dots$ by induction on the natural filtration of R, $R = \cup_n R(n)$ with $dR(0) \subset \wedge U \otimes \wedge W$, and for $n > 0$, $dR(n) \subset (\wedge U \otimes \wedge W) \otimes \wedge R(n-1)$. We begin with d_2. If $x \in R(0)$, then $d_2 d_1 x$ is a d_1-cycle in the kernel of ρ. Since ρ is a quasi-isomorphism for d_1, $d_2 d_1 x = d_1 \alpha$ with $\rho(\alpha) = 0$. We define $d_2 x = -\alpha$. This process can be extended to describe d_2, and then after d_3, and so on. This implies (iii).

(iv) This is a direct consequence of (ii). $\square$

Proposition 8.8 *Denote by L_X and L_Y the homotopy Lie algebras of connected spaces X and Y. Then the inclusions $X, Y \to X \vee Y$ induce a natural isomorphism*

$$L_{X \vee Y} \xrightarrow{\cong} L_X \,\widehat{\amalg}\, L_Y.$$

Proof Let $(\wedge T, d) \xrightarrow{\simeq} \wedge W \times_{\mathbb{Q}} \wedge Q$. Then, by Proposition 8.7,

$$(\wedge T, d_1) \xrightarrow{\simeq} (\wedge W, d_1) \times_{\mathbb{Q}} (\wedge Q, d_1).$$

Therefore for any quadratic Sullivan algebra $(\wedge V, d_1)$, the based homotopy classes of morphisms satisfy

$$[(\wedge V, d_1), (\wedge T, d_1)]_* = [(\wedge V, d_1), (\wedge W, d_1) \times_{\mathbb{Q}} (\wedge Q, d_1)]_*$$

$$= [(\wedge V, d_1), (\wedge W, d_1)]_* \times [(\wedge V, d_1), (\wedge Q, d_1)]_*.$$

Proposition 4.1 translates this equality to give

$$\mathcal{C}(L_T, L_V) = \mathcal{C}(L_W, L_V) \times \mathcal{C}(L_Q, L_V),$$

where $\mathcal{C}(-, -)$ denotes the set of morphisms in the category Enrich of enriched Lie algebras. This identifies $L_T = L_W \mathbin{\widehat{\amalg}} L_Q$. The naturality is a standard diagram chasing.

$\square$

Proposition 8.9 *With the hypotheses and notation above, if $f : L \to L'$ and $g : E \to E'$ are morphisms of enriched Lie algebras, then*

$$f \mathbin{\widehat{\amalg}} g : L \mathbin{\widehat{\amalg}} E \to L' \mathbin{\widehat{\amalg}} E'$$

is surjective if and only if f and g are surjective.

Proof To show that if f and g are surjective, then $f \mathbin{\widehat{\amalg}} g$ is surjective, it is sufficient to consider the case $g = id_E : E \xrightarrow{\;=\;} E$. In this case we have the row-exact commutative diagram

$$
\begin{array}{ccccccccc}
0 & \longrightarrow & L_R & \longrightarrow & L \mathbin{\widehat{\amalg}} E & \longrightarrow & L \times E & \longrightarrow & 0 \\
 & & \downarrow{\scriptstyle h} & & \downarrow{\scriptstyle f \widehat{\amalg} id_E} & & \downarrow{\scriptstyle f \times id_E} & & \\
0 & \longrightarrow & L_{R'} & \longrightarrow & L' \mathbin{\widehat{\amalg}} E & \longrightarrow & L' \times E & \longrightarrow & 0,
\end{array}
$$

and so it is also sufficient to show that h is surjective.

By Proposition 8.7(i), L_R and $L_{R'}$ are profree. Moreover,

$$L_R / L_R^{(2)} \cong d(\wedge U_E) \otimes \wedge^{\geq 1} U_L \quad \text{and} \quad L_{R'} / L_{R'}^{(2)} \cong d(\wedge U_E) \otimes \wedge^{\geq 1} U_{L'}.$$

Since $f : L \to L'$ is surjective, the induced map $U_L \to U_{L'}$ is surjective, and therefore $L_R \to L_{R'}$ is also surjective.

In the reverse direction note that the maps id_L and $E \to 0$ define a retraction $r :$ $L \mathbin{\widehat{\amalg}} E \to L$. The corresponding retraction $r' : L' \mathbin{\widehat{\amalg}} E' \to L'$ fits in the commutative diagram

$$
\begin{array}{ccc}
L\,\widehat{\amalg}\,E & \xrightarrow{\;f\,\widehat{\amalg}\,g\;} & L'\,\widehat{\amalg}\,E' \\
\Big\downarrow{\scriptstyle r} & & \Big\downarrow{\scriptstyle r'} \\
L & \xrightarrow{\;\;\;f\;\;\;} & L'.
\end{array}
$$

Therefore if $f\,\widehat{\amalg}\,g$ is surjective, so is f and, similarly, so is g. $\qquad\square$

Proposition 8.10 *Let L and L' be enriched Lie algebras, and then:*

(i) The kernel of the projection $L\,\widehat{\amalg}\,L' \to L \times L'$ is profree.

(ii) If L and L' are both profree, then $L\,\widehat{\amalg}\,L'$ is also profree.

(iii) If L is profree, then the kernel K of the projection $L\,\widehat{\amalg}\,L' \to L'$ is profree. More precisely, if $L = \overline{\mathbb{L}}_S$, then $K \cong \overline{\mathbb{L}}_{S\,\widehat{\otimes}\,\overline{UL'}}$.

Proof

(i) Apply Proposition 8.7 to identify the kernel as the profree Lie algebra L_R.

(ii) If L and L' are profree, then by Proposition 8.6 there are quasi-isomorphisms

$$
\wedge W \xrightarrow{\;\simeq\;} \mathbb{Q} \oplus S \quad \text{and} \quad \wedge W' \xrightarrow{\;\simeq\;} \mathbb{Q} \oplus S'
$$

with $S \cdot S = 0 = S' \cdot S'$ and vanishing differentials in S and S'. Thus $\wedge W \times_{\mathbb{Q}} \wedge W' \simeq \mathbb{Q} \oplus S \oplus S'$, and its homotopy Lie algebra, $L\,\widehat{\amalg}\,L'$, is profree by Proposition 8.6.

(iii) Let $\wedge T$ be the minimal Sullivan model of $\wedge V_L \times_{\mathbb{Q}} \wedge V_{L'}$, and let $\wedge V_{L'} \otimes \wedge U_{L'}$ be the acyclic closure of $\wedge V_{L'}$. Then the quadratic model $\wedge K$ associated with the kernel of the projection $L\,\widehat{\amalg}\,L' \to L'$ is quasi-isomorphic to $\wedge T \otimes_{\wedge V_{L'}} (\wedge V_{L'} \otimes \wedge U_{L'})$. Since $\wedge V_L$ is quasi-isomorphic to $\mathbb{Q} \oplus E$ with $E \cdot E = 0$ for some E,

$$
\wedge V_L \times_{\mathbb{Q}} \wedge V_{L'} \cong E \oplus \wedge V_{L'}.
$$

Therefore we have a quasi-isomorphism

$$
\wedge K \simeq \mathbb{Q} \oplus (E \otimes \wedge U_{L'})
$$

with trivial multiplication and differential.

Recall now that (Proposition 5.2) $(\wedge U_{L'})^{\vee} \cong \overline{UL'}$. We then have $K = \overline{\mathbb{L}}_Z$, with

$$
sZ = (E \otimes \wedge U_{L'})^{\vee} = E^{\vee}\,\widehat{\otimes}\,\overline{UL'} = s(S\,\widehat{\otimes}\,\overline{UL'}).
$$

$\qquad\square$

8.5   Wedges of Spheres

Proposition 8.11 *A minimal Sullivan algebra $\wedge V$ with finite type homology is the Sullivan model of a wedge of spheres, X, if and only if its homotopy Lie algebra, L, is profree.*

Proof A wedge of spheres has LS category ≤ 1; therefore its minimal Sullivan model, $\wedge V$, is quadratic, and its homotopy Lie algebra is profree.

In the reverse direction suppose that L_V is profree. Then by Proposition 8.6, $\wedge V \xrightarrow{\simeq} \mathbb{Q} \oplus (V \cap \ker d)$, and $L = \overline{\mathbb{L}}_T$ with $sT = (V \cap \ker d)^\vee$. Since $H(\wedge V, d)$ is finite type, T is finite type. Choose a linearly ordered basis $\{x_i\}$ of T, and set $n_i = \deg x_i$ and $\mathcal{S}_X = \{i\}$. Then set

$$X = \vee_{i \in \mathcal{S}_X} S_i^{n_i}.$$

By construction, $H_{\geq 1}(X; \mathbb{Q}) = T$.

Now let $\wedge W$ be the minimal Sullivan model of X. Since X is a wedge of spheres,

$$\wedge W \simeq (\mathbb{Q} \oplus H^{\geq 1}(X), 0) = (\mathbb{Q} \oplus T^\vee, 0) = (\mathbb{Q} \oplus H^{\geq 1}(\wedge V), 0) \simeq \wedge V.$$

Since $\wedge W$ and $\wedge V$ are minimal Sullivan algebras, it follows that $\wedge V \cong \wedge W$. Therefore $\wedge V$ is the minimal Sullivan model of X.

$\square$

These constructions provide an explicit description of the rationalization map

$$\pi_*(\widetilde{\varphi}) : \pi_*(Y) \to \pi_*(Y_\mathbb{Q})$$

when $Y = \vee_\sigma S^{n_\sigma}$ is a wedge of spheres of finite type.

Proposition 8.12 *With the hypotheses and notation above, denote by $\sigma \in \pi_{n_\sigma}(S^{n_\sigma})$ the homotopy class represented by $\mathrm{id}_{S^{n_\sigma}}$. Then the homotopy Lie algebra L_Y of Y can be written $\overline{\mathbb{L}}_T$ in which $\pi_*\widetilde{\varphi}$ maps the set $\{\sigma\}$ bijectively to a basis of T.*

Proof Here $L_Y = \overline{\mathbb{L}}_T$ with $sT = (V \cap \ker d_1)^\vee$ (Proposition 8.6). Because Y is a wedge of finite type,

$$(V \cap \ker d_1)^\vee = H^{\geq 1}(Y)^\vee = H_{\geq 1}(Y; \mathbb{Q}).$$

However, a basis of $H_{\geq 1}(Y; \mathbb{Q})$ is provided by the images of the homotopy classes σ under the Hurewicz homomorphism. Since the Hurewicz homomorphism is natural, the proposition follows from Proposition 8.3. $\qquad\square$

Remark Let Z be a graded vector space concentrated in degree 2 with an infinite countable basis $v_1, v_2, \ldots$, and let $\wedge V$ be the minimal model of $\mathbb{Q} \oplus Z$. Then $V^2 \cong Z$, and its homotopy Lie algebra is a profree Lie algebra $\overline{\mathbb{L}}_T$ with $sT \cong (V^2)^\vee$.

However $\wedge V$ is not the minimal model of a space S because otherwise $H^2(S) = V^2$ must be the dual of the vector space $H_2(S)$ which is not the case. In particular $\overline{\mathbb{L}}_T$ is not the homotopy Lie algebra of a space and in particular of a wedge of spheres.

Indeed a wedge X of an infinite number of two-dimensional spheres has a minimal Sullivan model of the form $(\wedge V, d)$ with $(\wedge V, d) \cong (\mathbb{Q} \oplus V^2, 0)$. Moreover $V^2 = H^2(X)$ is the dual of an infinite dimensional vector space W. Therefore its homotopy Lie algebra is $\overline{\mathbb{L}}_T$ with $sT = (V^2)^\vee = (W^\vee)^\vee$.

The Sullivan Rationalization, $X_\mathbb{Q}$, of a Space X 9

Throughout this section we consider a connected space X and its rationalization map $\widetilde{\varphi} : X \to X_\mathbb{Q}$. We give new information on $H^*(X_\mathbb{Q})$ and on the kernel of $\pi_*(\widetilde{\varphi})$. A particular attention will be given to the Gottlieb groups of X and $X_\mathbb{Q}$ and their relations.

Examples

1. When X is simply connected with finite type rational homology, then $X \to X_\mathbb{Q}$ is a rational homotopy equivalence: $H_*(X; \mathbb{Q}) \cong H_*(X_\mathbb{Q})$ [33].
2. The Sullivan rationalization of a wedge of spaces is not always the wedge of the rationalizations. For instance, the injection $(S^1)_\mathbb{Q} \vee (S^1)_\mathbb{Q} \to (S^1 \vee S^1)_\mathbb{Q}$ is not a homotopy equivalence because $H_2((S^1 \vee S^1)_\mathbb{Q}) \neq 0$ [54].
3. For $n \geq 2$, the natural injection $X = S^1_\mathbb{Q} \vee S^n_\mathbb{Q} \to X_\mathbb{Q} = (S^1 \vee S^n)_\mathbb{Q}$ is not a homotopy equivalence. This can be seen by looking at their universal covers. First of all the universal cover of X is a wedge indexed by $\mathbb{Q}$ of rational spheres, and $H_n(\widetilde{X})$ is a vector space with a countably infinite dimension.

   On the other hand the minimal Sullivan model of X $(\wedge V, d)$ satisfies $V = V^1 \oplus V^n \oplus V^{>n}$. Basis for V^1 and V^n are given, respectively, by x and the v_n, $n \geq 0$, with $dx = 0$, $dv_0 = 0$, and $dv_n = xv_{n-1}$ for $n > 0$. Since the spatial realization of the relative Sullivan model $(\wedge V^1, 0) \to (\wedge V, d) \to (\wedge V^{>1}, \overline{d})$ is a fibration, the universal cover Y of $X_\mathbb{Q}$ is the space $\langle \wedge V^{>1}, \overline{d} \rangle$. Therefore

$$H_n(Y) \cong \pi_n(Y) \cong (V^n)^\vee$$

   is a non-countably infinite vector space.

Proposition 9.1 *Let $(\wedge V, d)$ be a minimal Sullivan algebra. When V is a finite type graded vector space, then $\iota_{\wedge V} : \wedge V \to A_{PL}\langle \wedge V \rangle$ is a quasi-isomorphism.*

Proof We first prove that $(\wedge V, d)$ is the Sullivan minimal model of a nilpotent space X. Write $V = \cup_n V(v)$, with $V(0) = V^1$, $V(n) \subset V(n+1)$, and $d : V(n) \to \wedge V(n-1)$. Since V is finite type for each integer p, there is some n with $V^{\leq p} \subset V(n)$.

Since V^1 is finite dimensional, $G = \pi_1(\wedge V)$ is a rational nilpotent finitely generated group with classifying space BG, and $(\wedge V^1, d)$ is the Sullivan minimal model of BG. Now we construct a sequence of fibrations $X(n) \to X(n-1)$ with $X(0) = BG$ and related to the $\wedge V(n)$ by a sequence of quasi-isomorphism

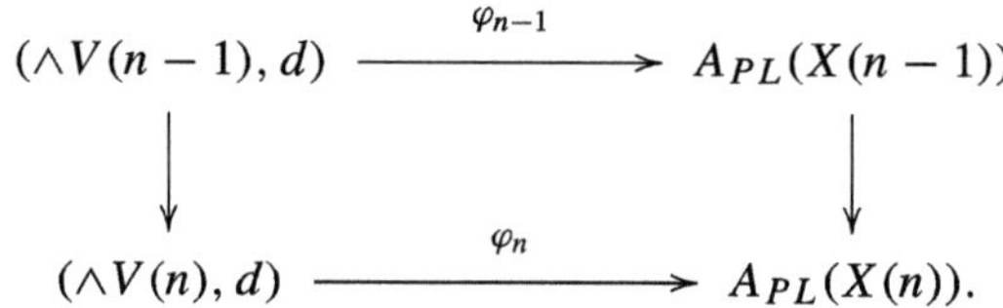

Suppose that φ_{n-1} has been defined; denote by $Z(n)$ a direct summand of $V(n-1)$ in $V(n)$ and by $\overline{Z(n)}$ the suspension of $Z(n)$. Then $BZ(n) = \prod_{r \geq 1} K(Z(n)^r, r)$, and $B\overline{Z(n)} = \prod_{r \geq 1} K(Z(n)^r, r+1)$. Let x_i be a basis of $Z(n)$. The elements dx_i are cycles in $\wedge Z(n-1)$. This induces a map $\tau : X(n-1) \to B\overline{Z(n)}$ associated with the classes $[\varphi_{n-1}(dx_i)]$.

We then construct the pullback fibration of the loop space fibration on $B\overline{Z(n)}$ along τ to get the diagram

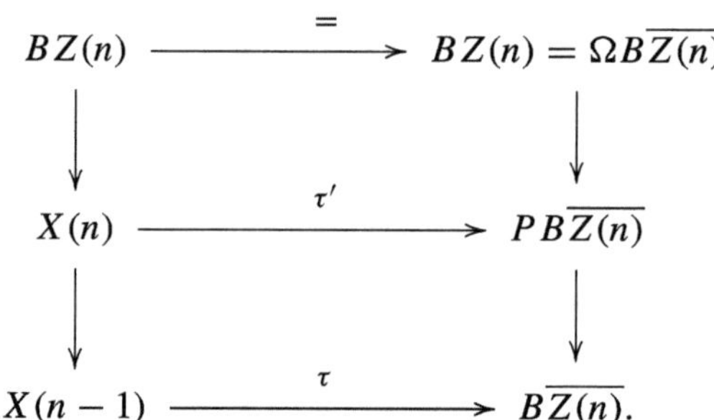

The commutativity of the induced diagram

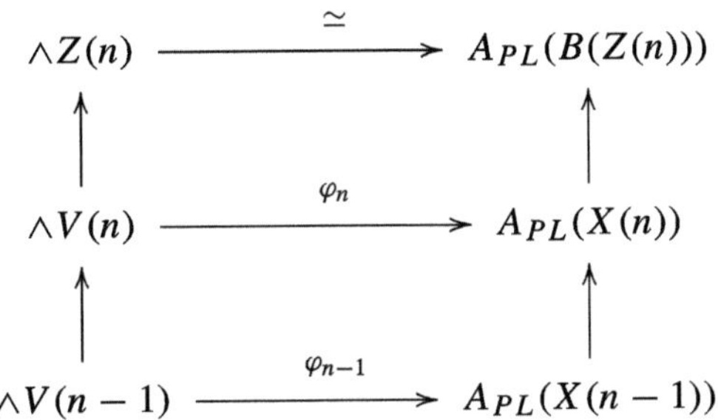

shows that φ_n is a quasi-isomorphism. We define then X to be the inverse limit of the tower $X(n)$. The morphisms φ_n fit together to give a morphism $\varphi : (\wedge V, d) \to A_{PL}(X)$ making commutative the diagram

$$
\begin{array}{ccc}
(\wedge V, d) & \xrightarrow{\ \varphi\ } & A_{PL}(X) \\
\downarrow & & \downarrow \\
(\wedge V(n), d) & \xrightarrow{\ \varphi_n\ } & A_{PL}(X(n))
\end{array}
$$

Since $H^{\leq n}(\varphi)$ is an isomorphism for all n, φ is a quasi-isomorphism.

It follows that $X_\mathbb{Q} \cong \langle \wedge V, d \rangle$. Since X is a nilpotent space with finite Betti numbers, the natural map $X \to X_\mathbb{Q}$ is a rational homotopy equivalence. Therefore X and $X_\mathbb{Q}$ have the same Sullivan minimal model, which implies the result.

$\square$

9.1   The Cohomology of $X_\mathbb{Q}$

In general $H(X_\mathbb{Q})$ remains mysterious, and it is unknown what conditions are necessary and or sufficient for $H^n(X_\mathbb{Q}; \mathbb{Z})$, $n \geq 1$, to be a rational vector space. We do have the following proposition:

Proposition 9.2

(i) *Suppose $\varphi : (\wedge V, d) \xrightarrow{\simeq} A_{PL}(X)$ is a minimal Sullivan model of a space X. Then $(\wedge V, d)$ is a retract of the minimal Sullivan model of $X_\mathbb{Q}$, and so $H(X)$ is a retract of $H(X_\mathbb{Q})$, and $X_\mathbb{Q}$ is a retract of $(X_\mathbb{Q})_\mathbb{Q}$.*

(ii) *Let $f : Y \to X$ be a continuous map between connected spaces, and let*

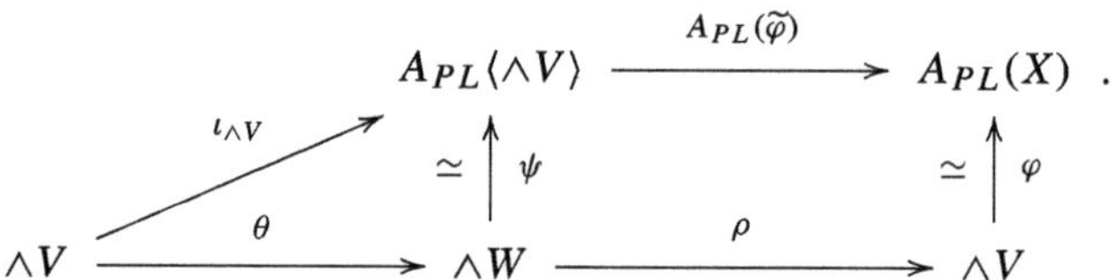

$$\begin{array}{ccc}
\wedge V & \xrightarrow{\ \varphi\ } & A_{PL}(X) \\
\downarrow{\scriptstyle\theta} & & \downarrow{\scriptstyle A_{PL}(f)} \\
\wedge V' & \xrightarrow{\ \psi\ } & A_{PL}(Y)
\end{array}$$

be a homotopy commutative diagram in which φ and ψ are minimal Sullivan models. Then θ is a homotopy retract of $A_{PL}\langle\theta\rangle$.

Proof

(i) Let $\psi : (\wedge W, d) \xrightarrow{\simeq} A_{PL}\langle\wedge V\rangle$ be the minimal Sullivan model of $X_{\mathbb{Q}}$. By the homotopy lifting property there is a map $\theta : \wedge V \to \wedge W$ such that $\psi \circ \theta \sim \iota_{\wedge V}$. A new application of the homotopy lifting property gives a map $\rho : \wedge W \to \wedge V$ such that $\varphi \circ \rho \sim A_{PL}(\widetilde{\varphi}) \circ \psi$. Thus we get a homotopy commutative diagram

$$\begin{array}{ccccc}
 & & A_{PL}\langle\wedge V\rangle & \xrightarrow{\ A_{PL}(\widetilde{\varphi})\ } & A_{PL}(X) \ . \\
 & \nearrow^{\iota_{\wedge V}} & {\scriptstyle\simeq}\uparrow{\scriptstyle\psi} & & {\scriptstyle\simeq}\uparrow{\scriptstyle\varphi} \\
\wedge V & \xrightarrow[\ \theta\ ]{} & \wedge W & \xrightarrow[\ \rho\ ]{} & \wedge V
\end{array}$$

By Lemma 7.1, $A_{PL}(\widetilde{\varphi}) \circ \iota_{\wedge V} \sim \varphi$. Therefore $\varphi\rho\theta \sim A_{PL}(\widetilde{\varphi}) \circ \iota_{\wedge V} \sim \varphi$, and since φ is a quasi-isomorphism, we have $\rho \circ \theta \sim id$.

(ii) By naturality we have the following homotopy commutative diagram which implies the result:

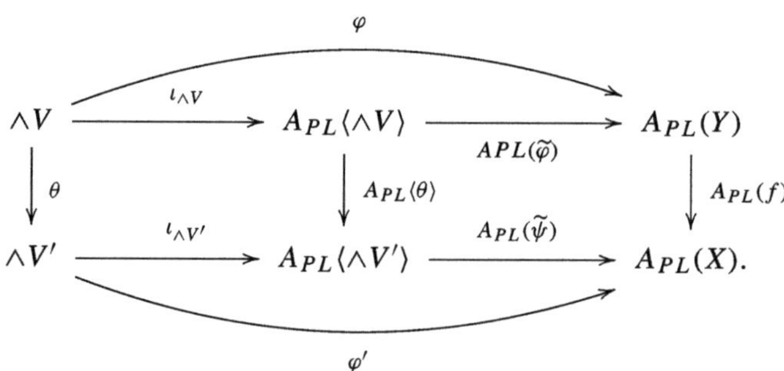

$$\begin{array}{ccccc}
\wedge V & \xrightarrow{\ \iota_{\wedge V}\ } & A_{PL}\langle\wedge V\rangle & \xrightarrow{\ A_{PL}(\widetilde{\varphi})\ } & A_{PL}(Y) \\
\downarrow{\scriptstyle\theta} & & \downarrow{\scriptstyle A_{PL}\langle\theta\rangle} & & \downarrow{\scriptstyle A_{PL}(f)} \\
\wedge V' & \xrightarrow{\ \iota_{\wedge V'}\ } & A_{PL}\langle\wedge V'\rangle & \xrightarrow{\ A_{PL}(\widetilde{\psi})\ } & A_{PL}(X).
\end{array}$$

$\square$

Corollary 9.1 *Any continuous map $f : Y \to X_{\mathbb{Q}}$ extends to a map $Y_{\mathbb{Q}} \to X_{\mathbb{Q}}$.*

Proof Compose $f_{\mathbb{Q}}$ with the retraction $(X_{\mathbb{Q}})_{\mathbb{Q}} \to X_{\mathbb{Q}}$. $\square$

Corollary 9.2 *The morphism*

$$[\wedge V, \wedge W] \to [\langle \wedge W \rangle, \langle \wedge V \rangle]$$

induced by spatial realizations admits a retraction and is thus injective. $\square$

Remark When $W = W^{>1}$ is a graded vector space of finite type, then $\iota_{\wedge W} : \wedge W \to A_{PL}\langle \wedge W \rangle$ is a homotopy equivalence, and classical obstruction theory shows that in that case the morphism $[\wedge V, \wedge W] \to [\langle \wedge W \rangle, \langle \wedge V \rangle]$ is a bijection [33, Theorem 17.12].

9.2   The Sullivan Rationalization and the Malcev Completion of a Group

We recall that for any group, G, G^n denotes the subgroup generated by iterated commutators of length n.

Definition If $f : BG \to (BG)_{\mathbb{Q}}$ is the Sullivan rationalization of the classifying space of a group, G, then the morphism

$$\pi_1(f) : G \to \pi_1(BG)_{\mathbb{Q}} := G_{\mathbb{Q}}$$

is the *Sullivan rationalization* of G.

Denote by $\varphi : \wedge V \to A_{PL}(BG)$ the minimal Sullivan model of BG. Then $f = \widetilde{\varphi} : BG \to \langle \wedge V \rangle$ is the adjoint map. Associated with $(\wedge V, d)$, we have three important constructions:

- The *homotopy Lie algebra, L*, of $(\wedge V, d)$, is an enriched Lie algebra, defined (Chap. 8) via the quadratic differential, d_1. By construction

$$sL = V^{\vee}$$

  as graded vector spaces.
- The *group G_L* (Sect. 3.1) connected with L by natural inverse bijections

$$L_0 \underset{\log}{\overset{\exp}{\rightleftarrows}} G_L.$$

- A *natural bijection of sets* (7.4)

$$\tau_V : \pi_1\langle\wedge V\rangle \xrightarrow{\;\cong\;} V^\vee.$$

Then [36, Theorem 2.4] provides a commutative diagram

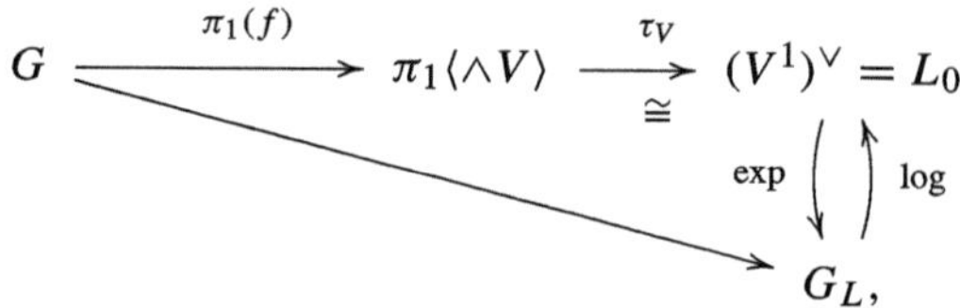

in which $\exp \circ \tau_V$ is a natural isomorphism of groups.

Thus (10.1) identifies the Sullivan rationalization $G \to G_{\mathbb{Q}}$ with the homomorphism

$$(\exp \circ \tau_V) \circ \pi_1(\widetilde\varphi) : G \to G_L.$$

Proposition 9.3 *Let $\varphi : \wedge V \xrightarrow{\;\cong\;} A_{PL}(BG)$ be the minimal Sullivan model of a classifying space. Suppose that $H^*(BG)$ is a graded vector space of finite type. Then the diagram*

$$
\begin{array}{ccc}
G & \xrightarrow{\;\tau_V \circ \pi_1(\widetilde\varphi)\;} & (V^1)^\vee \\[2mm]
\downarrow & & \downarrow{\scriptstyle j^\vee} \\[2mm]
G/G^2 \otimes \mathbb{Q} & \xrightarrow[\cong]{\;hur\;} & (V^1 \cap \ker d)^\vee
\end{array}
$$

is commutative.

Proof The isomorphism $V^1 \cap \ker d \xrightarrow{\;\cong\;} H^1(BG)$ dualizes to an isomorphism $H_1(BG;\mathbb{Q}) \xrightarrow{\;\cong\;} (V^1 \cap \ker d)^\vee$. It follows from Proposition 8.3 that the composite

$$G \to (V^1)^\vee \xrightarrow{\;j^\vee\;} (V^1 \cap \ker d)^\vee \xrightarrow{\;\cong\;} H_1(X;\mathbb{Q})$$

is $hur \otimes \mathbb{Q}$, where hur is the classical Hurewicz homomorphism. Thus it induces an isomorphism $G/G^2 \otimes \mathbb{Q} \xrightarrow{\;\cong\;} (V \cap \ker d_1)^\vee$. $\qquad\square$

For recall, the *Malcev completion of a finitely generated group G* is a morphism

$$\gamma : G \to \widehat{G}$$

satisfying the following properties:

(i) Each $\widehat{G}^n/\widehat{G}^{n+1}$ is a rational vector space.

(ii) The induced maps

$$\gamma_n : G^n/G^{n+1} \otimes \mathbb{Q} \xrightarrow{\;\cong\;} \widehat{G}^n/\widehat{G}^{n+1}$$

are isomorphisms.

(iii) $\widehat{G}$ is complete with respect to the quotients $\widehat{G}^n/\widehat{G}^{n+1}$,

$$\widehat{G} = \varprojlim_n \widehat{G}^n/\widehat{G}^{n+1}.$$

In particular, when $\dim G/G^2 \otimes \mathbb{Q}$ is finite [36, Theorem 7.5], then

$$G \to G_{\mathbb{Q}}$$

is the Malcev completion of G.

A direct construction of $\widehat{G}$ is described in [48]. It follows from the fundamental theorem [48, Page 7] that for each $x \in \widehat{G}^n/\widehat{G}^{n+1}$ there are an integer p and an element $y \in G^n$ such that $\gamma_n(y \otimes 1) = px$.

Example: Free Groups

Suppose $b_1, \ldots, b_s$ freely generate a free group G. The classifying space for G is then

$$X = S^1_{b_1} \vee \cdots \vee S^1_{b_s},$$

and b_i is the homotopy class of the inclusion of $S^1_{b_i}$ in X. In particular, $\pi_*(X) = \pi_1(X) = G$.

On the other hand, Corollary 3.2 to Proposition 8.6, the minimal Sullivan model of X satisfies the conditions of Proposition 8.6 with $S = H^1(X)$. Therefore it is a quadratic Sullivan algebra $(\wedge V, d_1)$, and the inclusion

$$j : V^1 \cap \ker d_1 \longrightarrow \wedge^{\geq 1} V$$

induces an isomorphism $\mathbb{Q} \oplus (V^1 \cap \ker d_1) \xrightarrow{\cong} H(\wedge V) = H(X)$. It follows that $V = V^1$. It follows that the enriched Lie algebra L_X is a profree Lie algebra.

Proposition 9.4

(i) The elements $\log_G b_i$ are the basis of a subspace $T \subset L_X$ for which $L_X = \overline{\mathbb{L}}_T$.

(ii) If $c, d \in G$, then

$$\log_G cd - (\log_G c + \log_G d + \frac{1}{2}[\log_G c, \log_G d]) \in [\log_G c, \log_G d] \cdot \bar{J},$$

and

$$\log_G(cdc^{-1}d^{-1}) - [\log_G c, \log_G d] \in [\log_G c, \log_G d] \cdot \bar{J}.$$

Proof

(i) The $[b_i]$ form a basis of $G/G^2 \otimes \mathbb{Q}$. Their images by $\log_G$ form thus a basis L_V/L_V^2. This gives the result.

(ii) A naturality argument reduces this to the case G is freely generated by c and d. In this case $T = \mathbb{Q}\log_G c \oplus \mathbb{Q}\log_G d$, $[T, T] = \mathbb{Q}[\log_G c, \log_G d]$, and $L^{(3)} \subset [\log_G c, \log_G d] \cdot \bar{J}$. Thus (ii) follows immediately from [36, Proposition 2.4]. $\square$

9.3   The Kernel of the Sullivan Rationalization $\pi_1(\widetilde{\varphi}) : \pi_1(X) \to \pi_1(X_{\mathbb{Q}})$

Let G be any group. The normal subgroup $Tor(G)$ is defined by

$$Tor(G) = \{a \in G \mid \text{for each } n \text{ there is an integer } k \text{ such that } a^k \in G^n\}.$$

(Thus if G is abelian, $Tor(G)$ is the subgroup of torsion elements.)

Example Let G be the group defined by the presentation $< a, b; bab^{-1}a^{-3} = 1 >$. Then for each integer n, $a^{2n} \in G^{n+1}$. First of all, $a^2 = (b, a)$. Then suppose that $a^{2n} \in G^{n+1}$; then $[a^{2n}, b] = a^{2n}ba^{-2n}b^{-1} = a^{2n}(ba^{-1}b^{-1})^{2n} = a^{-4n}$. Therefore $a \in Tor(G)$.

Proposition 9.5 *If $H_1(X; \mathbb{Z})$ is finitely generated, then*

$$ker\,(\pi_1(\widetilde{\varphi}) : \pi_1(X) \to \pi_1(X_{\mathbb{Q}})) = Tor(\pi_1(X)).$$

Proof For simplicity, we denote $\pi_1(X)$ by G. Then let $(\wedge V, d)$ and $(\wedge W, d)$ be the minimal Sullivan models of X and BG. The classifying map $X \to BG$ induces an isomorphism in $H_1(-; \mathbb{Z})$ and, by a theorem of H. Hopf [13, Theorem 5.2], a surjection in $H_2(-; \mathbb{Q})$. Therefore it induces an isomorphism in $H^1(-; \mathbb{Q})$ and an injection in $H^2(-; \mathbb{Q})$. In particular the corresponding morphism of minimal Sullivan models restricts to an isomorphism

$$(\wedge V^1, d) \xleftarrow{\ \cong\ } (\wedge W^1, d).$$

This identifies $\pi_1(X_{\mathbb{Q}})$ with $\pi_1((BG)_{\mathbb{Q}})$ and reduces the proposition to the case $X = BG$.

Denote by L the component of degree 0 in the homotopy Lie algebra of BG. Since (by hypothesis) G/G^2 is finitely generated, it follows that each n, the group G^n/G^{n+1} is finitely generated. In particular,

$$\ker\left(G^{n-1}/G^n \to \left(G^{n-1}/G^n\right) \otimes_{\mathbb{Z}} \mathbb{Q}\right)$$

is a torsion group. On the other hand, $\log \circ \pi_1(\widetilde{\varphi})$ induces isomorphisms [36, Theorem 7.5]

$$\psi_n : \left(G^{n-1}/G^n\right) \otimes_{\mathbb{Z}} \mathbb{Q} \xrightarrow{\cong} L^{n-1}/L^n.$$

Now suppose $a \in \ker \pi_1(\widetilde{\varphi})$ and that $a^k \in G^{n-1}$. Let b be the image of a^k in G^{n-1}/G^n. Then $\psi_n b$ is the image of $\log \pi_1(\widetilde{\varphi})a^k = k \log \pi_1(\widetilde{\varphi})a = 0$. Thus b is in the kernel of $G^{n-1}/G^n \to G^{n-1}/G^n \otimes_{\mathbb{Z}} \mathbb{Q}$, and so for some integer ℓ,

$$a^{\ell k} \in G^n.$$

It follows by induction that a is in $Tor(G)$.

In the reverse direction the same argument shows that if $a \in Tor(G)$, then for each n and some k,

$$k \log \pi_1(\widetilde{\varphi})a = \log \pi_1(\widetilde{\varphi})a^k \in L^n.$$

It follows that $\log \pi_1(\widetilde{\varphi})a \in \cap_n L^n = 0$, and thus $a \in \ker \pi_1(\widetilde{\varphi})$. $\square$

9.4   The Kernel of $\pi_*(\widetilde{\varphi})$ when X Is Simply Connected

In this subsection we suppose that X is a 1-connected CW complex, and we denote by

$$\rho_* = \pi_*(\widetilde{\varphi}) \otimes \mathbb{Q} : \pi_*(X) \otimes \mathbb{Q} \to \pi_*(X_{\mathbb{Q}})$$

the morphism induced by the rationalization. If X is a finite CW complex, a classical theorem of Sullivan [36, Theorem 1.6] asserts that ρ_* is an isomorphism. Since X is the union of its finite 1-connected subcomplexes, this yields the commutative diagram

$$
\begin{array}{ccc}
\varinjlim_\alpha \pi_*(X_\alpha) \otimes \mathbb{Q} & \xrightarrow{\ \cong\ } & \pi_*(X) \otimes \mathbb{Q} \\
\Big\downarrow{\scriptstyle \cong} & & \Big\downarrow \\
\varinjlim_\alpha \pi_*(X_\alpha)_{\mathbb{Q}} & \longrightarrow & \pi_*(X_{\mathbb{Q}}).
\end{array}
$$

On the other hand, let $\wedge V$ and $\wedge V_\alpha$ be, respectively, the minimal Sullivan models of X and X_α. Then

$$X_{\mathbb{Q}} = \langle \wedge V \rangle = \mathrm{Cdga}(\wedge V, A_{PL}) = \varprojlim_\alpha \mathrm{Cdga}(\wedge V_\alpha, A_{PL}) = \varprojlim_\alpha \langle \wedge V_\alpha \rangle.$$

Proposition 9.6 *An element a is in the kernel of ρ_n if and only if for each map $f : X \to Y$ with image is a 1-connected finite type CW complex; we have $(\pi_n(f) \otimes \mathbb{Q})(a) = 0$.*

Proof Suppose first that $a \in \ker \rho_n$, and let $f : X \to Y$ be a continuous map with Y a 1-connected finite type CW complex. Then we have a commutative diagram

$$
\begin{array}{ccc}
X_{\mathbb{Q}} & \xrightarrow{\ f_{\mathbb{Q}}\ } & Y_{\mathbb{Q}} \\[4pt]
{\scriptstyle \widetilde{\varphi}_X}\big\uparrow & & {\scriptstyle \widetilde{\varphi}_Y}\big\uparrow \\[4pt]
X & \xrightarrow{\ f\ } & Y.
\end{array}
$$

The commutativity of the diagram implies that

$$(\pi_n(\widetilde{\varphi}_Y) \otimes \mathbb{Q} \circ \pi_n(f) \otimes \mathbb{Q})(a) = 0.$$

Since Y is a 1-connected finite type CW complex, $\pi_*(\widetilde{\varphi}_Y) \otimes \mathbb{Q}$ is an isomorphism, and therefore $(\pi_n(f) \otimes \mathbb{Q})(a) = 0$.

Conversely, suppose that the image of a is zero in $\pi_n(Y) \otimes \mathbb{Q}$ for all maps $f : X \to Y$ when Y is a simply connected finite type CW complex. Denote by p_α the projection $\langle \wedge V \rangle \to \langle \wedge V_\alpha \rangle$. Since $\langle \wedge V_\alpha \rangle$ is a simply connected space with finite type homology, the composite

$$
X \xrightarrow{\ \widetilde{\varphi}\ } \langle \wedge V \rangle \xrightarrow{\ p_\alpha\ } \langle \wedge V_\alpha \rangle
$$

factorizes up to homotopy through some 1-connected CW complex, Z, of finite type. It follows that $\pi_*(p_\alpha \circ \widetilde{\varphi})(a) = 0$. Now since $\pi_*\langle \wedge V \rangle = \varprojlim_\alpha \pi_*\langle \wedge V_\alpha \rangle$, it follows that $\rho_n(a) = 0$.

$\square$

Example We end this section with an example of a space X for which ρ_n is not injective.

Let $Y = \vee_{i=1}^{\infty} S_i^3$. We denote by $a_i \in \pi_3(Y)$ the class of a homeomorphism $S^3 \to S_i^3 \subset Y$, and $e_i = hur(a_i) \in H_3(Y)$. Then $\pi_3(Y) \otimes \mathbb{Q} = \oplus_{i=1}^{\infty} \mathbb{Q}a_i$ and $H_3(Y) = \oplus_{i=1}^{\infty} \mathbb{Q}e_i$. Now let $x_i \in H^3(Y)$ be the class defined by

$$\langle x_i, e_j \rangle = \delta_{ij}.$$

The space $H^3(Y)$ is then isomorphic to the space of series $\sum_{i\geq 1} \alpha_i x_i$, $\alpha_i \in \mathbb{Q}$.

Now let X be the space obtained from Y by adding six-dimensional cells along the elements

$$[a_i, a_j] \quad \text{for } j > i + 1$$

and the elements

$$[a_{2i-1}, a_{2i}] - [a_1, a_2], \, i > 1.$$

The space Y is the increasing union of the spaces $Y_n = \vee_{i=1}^{2n} S_i^3$, and X is the increasing union of six-dimensional CW complexes X_n obtained from Y_n by adding cells to Y_n.

Of particular interest is the element $[a_1, a_2]$. Its image in $\pi_5(X)$ is nonzero because it is nonzero in each $\pi_5(X_n)$ (which can be verified because it is nonzero in $\pi_5((X_n)_\mathbb{Q})$). Now we prove that $\rho_5[a_1, a_2] = 0$.

A simple computation shows that the cohomology of X_n is given by

$$H^*(X_n) = \wedge(x_1, \ldots, x_{2n})/I_n$$

where $I_n = \mathbb{Q}\omega_n \oplus \wedge^{\geq 3}(x_i)$, with $\omega_n = \sum_{i=1}^n x_{2i-1}x_{2i}$. In particular the minimal Sullivan model of X_n, $(\wedge V(n), d)$, satisfies

$$V(n)^3 = \mathbb{Q}(x_1, \ldots, x_{2n}),$$

$$V(n)^5 = \mathbb{Q}t, \qquad dt = \omega_n.$$

Remark that $Y_1 = X_1 = S_1^3 \vee S_2^3$ and that the minimal Sullivan model of the injection $X_1 \to X_n$, $\rho : (\wedge V(n), d) \to (\wedge V(1), d)$ satisfies $\rho(x_1) = x_1$, $\rho(x_2) = x_2$, $\rho(x_i) = 0$ for $i > 2$ and $\rho(t) = t$.

Now denote by $(\wedge V, d)$ the minimal Sullivan model of X. We can identify V^3 with $H^3(Y)$. Let $\varphi_n : (\wedge V, d) \to (\wedge V(n), d)$ be the Sullivan representative of the injection $X_n \to X$. By construction, $\varphi_n(x_i) = x_i$, $i \leq 2n$, $\varphi_n(x_i) = 0$, $i > 2n$.

We claim that for any n there is no generator $u \in V^5$ such that $\varphi_n(u) = t$. This implies that the image of $[a_1, a_2]$ is zero in $\pi_5(X_\mathbb{Q})$, and so ρ_5 is not injective. Thus, we suppose that there is some $u \in V^5$ with $\varphi_n(u) = t$ and arrive to a contradiction. Note that by construction, $du \in \wedge^2 V^3$.

First, consider the vector space $\wedge^2 V(n)$, and set

$$E = \varprojlim_n \wedge^2 V(n).$$

A particularly interesting element in E is the element

$$\omega = \sum_{i=1}^{\infty} x_{2i-1} x_{2i}.$$

On the other hand multiplication induces a morphism $\gamma : \wedge^2 V^3 \to E$, and we show that ω is not in the image of γ. This implies that for some r, $du - \omega \neq 0$ in $\wedge(x_1, \ldots, x_{2r})$. In particular, $\varphi_r(u) \neq t$.

To show that $\omega \notin \operatorname{Im} \gamma$, to each $t \in V^3$ we associate the line vector t_i of its components

$$(t_i) = (\langle t, e_1 \rangle, \langle t, e_2 \rangle, \ldots).$$

To each product $tt' \in \wedge^2 V^3$ we associate the antisymmetric matrix $(t_i)^t \cdot t'_j - (t'_i)^t \cdot t_j$.

To an element $\sum_{i<j} \alpha_{ij} x_i x_j$ of $\wedge^2 V^3$ is associated with the antisymmetric matrix A with for $i > j$, $A_{ij} = \alpha_{ij}$. By construction the matrix associated with $\gamma(tt')$ is the matrix $(t_i)^t \cdot t'_j - (t'_i)^t \cdot t_j$.

Then remark that the rank of the matrix $\overline{t_i} \cdot t'_j$ is one, so that the rank of any element of $\wedge^2 V^3$ is finite. On the other hand, the matrix associated with ω is a block matrix composed of matrices of the form

$$\begin{pmatrix} 0 & 1 \\ -1 & 0 \end{pmatrix}.$$

Since it contains arbitrary large matrices with nonzero determinant, its rank is infinite. This proves that $\omega \notin \wedge^2 V^3$. $\square$

9.5   The Radical and Center of $\pi_*(X_{\mathbb{Q}})$

As recalled in Chap. 6, the category, $\operatorname{cat}(\wedge V)$, is the least m (or ∞) for which $\wedge V$ is a homotopy retract of $\wedge V / \wedge^{>m} V$. Moreover [36, Theorem 9.2]

$$\operatorname{cat}(\wedge V) \leq \operatorname{cat} X,$$

where $\operatorname{cat} X$ is the classical Lusternik-Schnirelmann category of X. On the other hand, the *radical*, $\operatorname{rad} L_V$, of L_V is the sum of the solvable ideals of L_V. Thus combining [27, Theorem 1] with [25, Theorem C] yields the following proposition:

Proposition 9.7 *Suppose $\operatorname{cat}(X) < \infty$. Then:*

(i) The sum of the solvable ideals, $\operatorname{rad} L_V$, is finite dimensional and

$$\dim(\operatorname{rad}(L_V)_{even}) \leq \operatorname{cat} X.$$

(ii) The union of the normal solvable subgroups of $\pi_1(X_{\mathbb{Q}})$ is a normal solvable subgroup of solvable length $\leq cat\,(X)$.

Proof (i) is immediate from the remarks above.

For the proof of (ii) note that $\mathrm{rad}(L_V)_0$ is a closed ideal since it is finite dimensional. Thus $\exp(\mathrm{rad}(L_V)_0)$ is a normal subgroup of $\pi_1(X_{\mathbb{Q}})$ whose quotient is $\exp((L_V)_0/\mathrm{rad}(L_V)_0)$. Therefore, since the center of $(L_V)_0/\mathrm{rad}(L_V)_0$ is zero, $\exp(L_V/\mathrm{rad}(L_V)_0)$ has no center and therefore no normal solvable subgroup. $\square$

Now define $C_*(X_{\mathbb{Q}}) \subset \pi_*(X_{\mathbb{Q}})$ to be the center of the subgroup $\pi_*(X_{\mathbb{Q}})$. It follows from Proposition 8.2 that $C_*(X_{\mathbb{Q}})$ corresponds to the center of the homotopy Lie algebra L_V. Thus Proposition 9.7 has the following Corollary.

Corollary *If $cat\,X < \infty$, then the group $C_*(X_{\mathbb{Q}})$ is a finite dimensional rational vector space.*

9.6   The Gottlieb Groups $G_*(X)$ and $G_*(X_{\mathbb{Q}})$

The nth *Gottlieb group*, $G_n(X)$, is the subgroup of $\pi_n(X)$ formed by the homotopy classes $[f]$ of maps $f : S^n \to X$ such that $f \vee \mathrm{id} : S^n \vee X \to X$ extends to the product $S^n \times X$. Since $[G_*(X), \pi_*(X)]$ is zero, $G_*(X) \subset C_*(X)$ and $G_*(X_{\mathbb{Q}}) \subset C_*(X_{\mathbb{Q}})$.

More generally, if $g : Y \to X$ is a continuous map, then the nth *Gottlieb group of g*, $G_n(g)$, is the subgroup of $\pi_n(X)$ formed by the homotopy classes of maps $f : S^n \to X$ such that the map $f \vee g : S^n \vee Y \to X$ extends to $S^n \times Y$.

Finally, recall from (7.3) that for any minimal Sullivan algebra $\wedge W$,

$$\pi_n\langle\wedge W\rangle = [\wedge W, \mathbb{Q} \oplus \mathbb{Q}a_n]$$

where a_n is the fundamental cohomology class in $H^n(S^n)$. We define the *Gottlieb group*, $G_n(\wedge W) \subset \pi_n\langle\wedge W\rangle$, to be the subgroup formed by those elements represented by morphisms $\sigma : \wedge W \to \mathbb{Q} \oplus \mathbb{Q}a_n$ such that $\sigma \times \mathrm{id} : \wedge W \to (\mathbb{Q} \oplus \mathbb{Q}a_n) \times_{\mathbb{Q}} \wedge W$ factors as

$$\wedge W \to (\mathbb{Q} \oplus \mathbb{Q}a_n) \otimes \wedge W \to (\mathbb{Q} \oplus \mathbb{Q}a_n) \times_{\mathbb{Q}} \wedge W.$$

Note here that

$$(\mathbb{Q} \oplus \mathbb{Q}a_n) \times_{\mathbb{Q}} \wedge W = \mathbb{Q}a_n \oplus \wedge W$$

and that σ is just the map

$$W = \wedge^{\geq 1} W / \wedge^{\geq 2} W \to \mathbb{Q}a_n.$$

Remark also that by definition $G_*(\wedge W)$ is contained in the center of the homotopy Lie algebra L_W of $\wedge W$.

Lemma 9.1 $\pi_*(\widetilde{\varphi}) : G_*(X) \to G_*(X_{\mathbb{Q}})$.

Proof If $f : S^n \to X$ represents an element of $G_*(X)$, we may extend $f \vee id_X$ to $g : S^n \times X \to X$. Now Proposition 8.4(iii) identifies $(S^n \times X)_{\mathbb{Q}} = S^n_{\mathbb{Q}} \times X_{\mathbb{Q}}$. Thus we obtain $g_{\mathbb{Q}} : S^n_{\mathbb{Q}} \times X_{\mathbb{Q}} \to X_{\mathbb{Q}}$. It follows that $g_{\mathbb{Q}}$ restricts to $\widetilde{\varphi} \circ f : S^n \to X_{\mathbb{Q}}$. Thus the composition $g_{\mathbb{Q}} : S^n \times X_{\mathbb{Q}} \to X_{\mathbb{Q}}$ extends $(\widetilde{\varphi} \circ f) \vee id_{X_{\mathbb{Q}}}$.

$\square$

Proposition 9.8 *Let* $\varphi : (\wedge V, d) \to A_{PL}(X)$ *be the minimal Sullivan model of a path connected space X and $\widetilde{\varphi} : X \to X_{\mathbb{Q}}$ the associated rationalization. Then*

$$G_*(\wedge V) = G_*(\widetilde{\varphi}) = G_*(X_{\mathbb{Q}}).$$

Proof $G_*(\wedge V)$, $G_*(\widetilde{\varphi})$, and $G_*(X_{\mathbb{Q}})$ are three subgroups of $\pi_*\langle \wedge V \rangle = \pi_*(X_{\mathbb{Q}})$. We proceed in three steps.

 Step 1. $G_*(\wedge V) \subset G_*(\widetilde{\varphi})$.
   Let $\wedge Z$ denote the minimal model of S^n; then (7.3)

$$\pi_n\langle \wedge V \rangle = \operatorname{Hom}(V^n, \mathbb{Q}) = [\wedge V, \wedge Z] = [\wedge V, A_{PL}(S^n)].$$

Thus an element f in $G_n(\wedge V)$ is represented by a morphism $\sigma : \wedge V \to \wedge Z$ for which $\sigma \times_{\mathbb{Q}} id : \wedge V \to \wedge Z \times_{\mathbb{Q}} \wedge V$ factors up to homotopy as

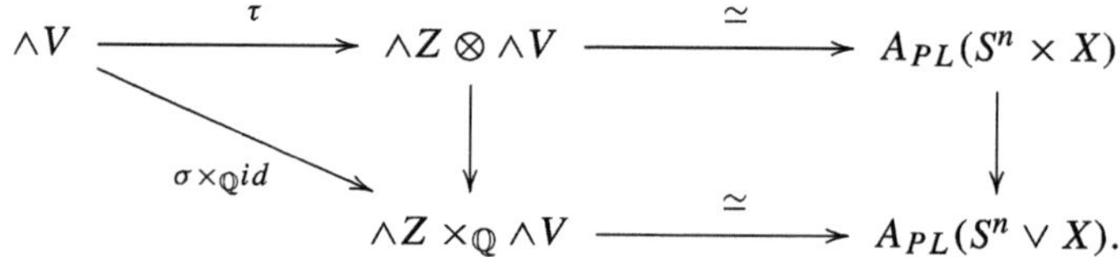

By adjonction this gives the diagram

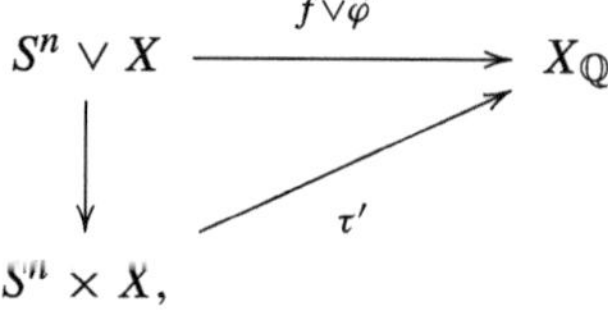

where τ' is the adjoint of τ. This proves that $[f] \in G_*(\widetilde{\varphi})$.

Step 2. $G_*(\widetilde{\varphi}) \subset G_*(X_\mathbb{Q})$.

Suppose $[f] \in G_*(\widetilde{\varphi})$. Then apply rationalization to the diagram

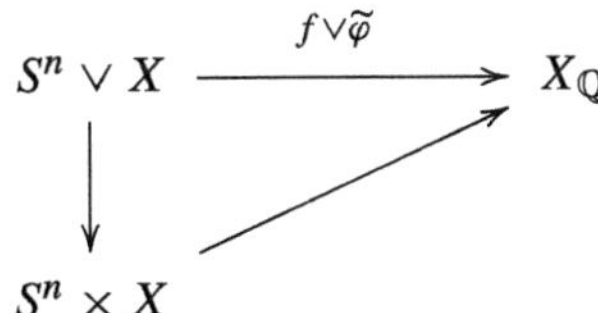

and use the fact that $X_\mathbb{Q}$ is a retract of $(X_\mathbb{Q})_\mathbb{Q}$ to get the result.

Step 3. $G_*(X_\mathbb{Q}) \subset G_*(\wedge V)$.

Let $f \in G_n(X_\mathbb{Q})$. We simply remark that the adjonction transforms the diagram

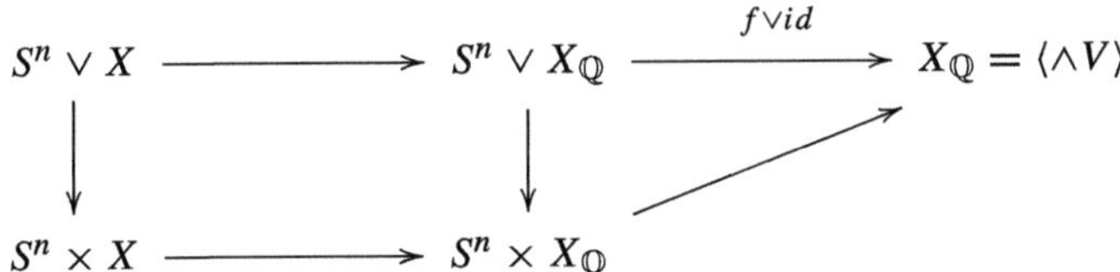

into the commutative diagram

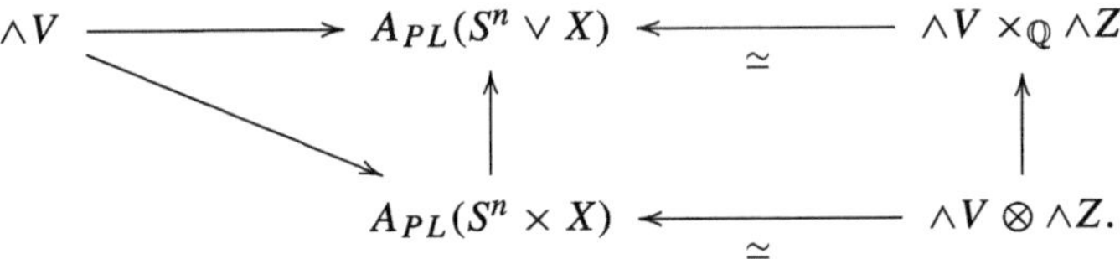

This implies the result.

$\square$

Recall next [36, §3.3] that a *minimal Sullivan extension* is a morphism

$$\wedge W \to \wedge W \otimes \wedge V, \qquad w \mapsto w \otimes 1,$$

of Sullivan algebras in which $\wedge W$ and the quotient $(\wedge V, d_V) = \mathbb{Q} \otimes_{\wedge W} (\wedge W \otimes \wedge V)$ are minimal Sullivan algebras. However the differential d in $\wedge W \otimes \wedge V$ may have a component d_0 of wedge degree 0: $d_0 : V \to W$. The dual of d_0 is then a linear map

$$\delta : W^\vee \to V^\vee$$

of degree -1.

On the other hand the geometric realization of the extension is a fibration

$$\langle \wedge V \rangle \to \langle \wedge W \otimes \wedge V \rangle \to \langle \wedge W \rangle$$

with fiber $F = \langle \wedge V \rangle$ and basis $\langle \wedge W \rangle$. The component $\delta : W^{\vee} \to V^{\vee}$ is the connecting map of the fibration

$$\delta : \pi_*\langle \wedge W \rangle \to \pi_{*-1}\langle \wedge V \rangle.$$

Proposition 9.9 *With the notation and hypotheses above*

$$Im\,\delta \subset G_*(\wedge V).$$

Proof Fix n and choose a cdga surjection

$$\xi : (\wedge W, d) \to (\mathbb{Q} \oplus d_0 V^n, 0)$$

with $\wedge^{\geq 2} W \subset \ker \xi$. Then extend $(\mathbb{Q} \oplus d_0 V^n, 0)$ to an acyclic cdga

$$(\mathbb{Q} \oplus d_0 V^n) \otimes \wedge U \xrightarrow{\simeq} \mathbb{Q}$$

with $U = U^{\geq n}$, and

$$d : U^n \xrightarrow{\cong} d_0 V^n \quad \text{and } d : U^{>n} \to \wedge^{\geq 1} U \otimes d_0 V^n.$$

Now using ξ we form the cdga

$$(\mathbb{Q} \oplus d_0 V^n) \otimes \wedge V := (\mathbb{Q} \oplus d_0 V^n) \otimes_{\wedge W} (\wedge W \otimes \wedge V).$$

This induces a surjective quasi-isomorphism

$$\rho : (\mathbb{Q} \oplus d_0 V^n) \otimes \wedge V \otimes \wedge U \xrightarrow{\simeq} (\wedge V, d_V).$$

Thus ρ has a left inverse $\sigma : (\wedge V, d_V) \to (\mathbb{Q} \oplus d_0 V^n) \otimes \wedge V \otimes \wedge U$, and it is immediate from the definitions that σ may be chosen so that $\sigma v = v$, $v \in V^{<n}$, and for $v \in V^n$,

$$\sigma v = v - \alpha(d_0 v),$$

where α is inverse to the isomorphism $U^n \xrightarrow{\cong} d_0 V^n$. Division by $d_0 V^n$ and by $U^{>n}$ then yields the sequence of morphisms

$$(\wedge V, d_V) \to (\wedge U^n, 0) \otimes (\wedge V, d_V) \to (\wedge U^n, 0) \times_{\mathbb{Q}} (\wedge V, d_V).$$

In particular, any morphism $\wedge U^n \to H(S^n)$ induces a sequence

$$(\wedge V, d_V) \to H(S^n) \otimes (\wedge V, d_V) \to H(S^n) \times_{\mathbb{Q}} (\wedge V, d_V).$$

In particular, the resulting morphism $(\wedge V, d_V) \to H(S^n)$ is an element of $G_n(\wedge V)$.

Finally, this element of $G_n(\wedge V)$ is determined by the composite $V^n \to U^n \to H^n(S^n)$, and by construction the map $V^n \to U^n$ is given by $v \mapsto -\alpha(d_0 v)$. This gives the commutative triangles

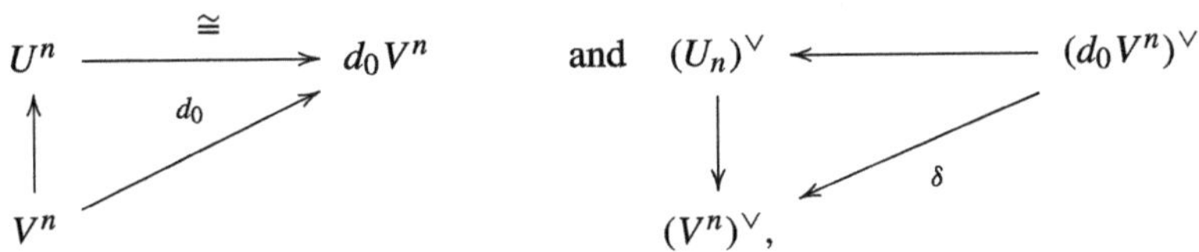

This identifies the elements in $\operatorname{Im} \delta$ as elements in $G_n(\wedge V)$.

$\square$

Corollary *Let $\wedge W \otimes \wedge V$ be a Sullivan extension. If $L_V = \overline{\mathbb{L}}_T$ is a profree Lie algebra with $\dim T \geq 2$, then the extension is minimal.*

Proof By Proposition 9.9, $\operatorname{Im} \delta \subset G_*(\wedge V)$, with $\delta = d_0^{\vee}$. Since $G_*(\wedge V)$ is in the center of L_V and $\dim T \geq 2$, the center of L_V is 0, and $d_0 = 0$. $\square$

Proposition 9.10 *If $\wedge V$ is a minimal Sullivan algebra and $\operatorname{cat}(\wedge V) < \infty$, then:*

(i) $G_{2n}(\wedge V) = 0$, $n \geq 1$.
(ii) $\dim G_*(\wedge V) \leq \operatorname{cat}(\wedge V)$.

Proof

(i) Let $B = \wedge a/(a^2)$, with $\deg a = 2n$. We suppose that there exists a morphism $\varphi : \wedge V \to B$ that is a representative of an element of $G_{2n}(\wedge V)$. If $\varphi \neq 0$, there is an element $z_0 \in V^{2n}$ with $\varphi(z_0) = a$.

Now let $(\wedge W, d) := \mathbb{Q} \otimes_{\wedge V^{<2n}} (\wedge V, d)$. By [36, Proposition 9.3] $\operatorname{cat}(\wedge W) \leq \operatorname{cat}(\wedge V) < \infty$. We denote by $\overline{\varphi} : \wedge W \to B$ the map induced by φ. Since $\overline{\varphi}$ is also in $G_{2n}(\wedge W)$, we get a morphism

$$\psi : \wedge W \to \wedge W \otimes B$$

with $\psi(w) - w \in \wedge W \otimes B^+$ and $\psi(z_0) = z_0 + a$. Thus $\psi(z_0^q) = (z_0 + a)^q$. By construction $d(z_0) = 0$. Since $\operatorname{cat} \wedge W < \infty$, there is $y \in \wedge W$ with $dy = z_0^q$. It follows that $\psi(z_0^q)$ is a boundary which is impossible because, by projection in B, a^q is not a boundary.

(ii) Suppose $G_{<n}(\wedge V) = 0$, and let $x \in G_n(\wedge V)$. We then choose an element $v_0 \in V^n$ such that $\langle v_0, sx \rangle = 0$ and denote $S = \{w \in V^n \,|, < w, sx > = 0\}$. Now let $Z = V/(V^{<n} \oplus S)$; this gives a surjective map of minimal Sullivan models $\wedge V \to \wedge Z$. Since $\mathrm{cat}(\wedge Z) \leq \mathrm{cat}(\wedge V)$ [36, Proposition 9.3], it is sufficient to show that $\dim G_*(\wedge Z) \leq \mathrm{cat}(\wedge Z)$. In other words, without loss of generality we may suppose $V = V^{\geq n}$ and that $x \in G_n(\wedge V)$, and v_0 is a cycle in V^n. Since (i) implies that n is odd, we also have

$$H(S^n) = \wedge a.$$

Moreover, since $x \in G_n(\wedge V)$, by definition we get a morphism

$$\varphi : \wedge V \longrightarrow \wedge a \otimes \wedge V, \qquad \text{with } \varphi(v_0) = a + v_0.$$

Moreover for each $v \in V$, $\varphi v - v \in \wedge^+ a \otimes \wedge V$.

Now $Z = \cup_{k \geq 1} W(k)$ with $W(1) = W \cap \ker d$ and $W(k+1) = W \cap d^{-1}(\wedge v_0 \otimes \wedge W(k))$. Suppose next, by induction on k, that $\wedge W(k)$ is preserved by d and that

$$\varphi w - w \in a v_0 \otimes \wedge W(k), \qquad w \in W(k).$$

Finally, fix a basis element z in a direct summand of $W(k)$ in $W(k+1)$. Then by definition,

$$dz = v_0 \Psi_1 + \Psi_2 \quad \text{and} \quad \varphi z = a(v_0 \Phi_1 + \Phi_2) + z,$$

where $\Phi_1, \Phi_2, \Psi_1,$ and $\Psi_2 \in \wedge W(k)$. This yields

$$d\varphi z = a v_0 d\Phi_1 - a d\Phi_2 + dz$$

$$\varphi dz = (a + v_0)\varphi \Psi_1 + \varphi \Psi_2 = \varphi \Psi_2.$$

Therefore $d\Phi_2 = -\Psi_1$.

Now extend the splitting to $W(k+1)$ by replacing z by $z - v_0 \Phi_2$. It is straightforward to check that our conditions are satisfied. Thus by induction we may write $\wedge V = \wedge v_0 \otimes \wedge W$ with $\wedge W$ preserved by d. It follows that $G_*(\wedge V) = \mathbb{Q}x \oplus G_*(\wedge W)$. Since by [36, Corollary 9.3] $\mathrm{cat}(\wedge W) = \mathrm{cat}(\wedge V) - 1$, (ii) follows by induction on $\mathrm{cat}(\wedge V)$. $\square$

Sullivan Rational Spaces

10

A Sullivan rational space is a space X for which the rationalization map $\widetilde{\varphi} : X \to X_{\mathbb{Q}}$ is a homotopy equivalence. For instance, the rationalization of a finite type simply connected space is a Sullivan rational space. The main question is to know if this is the only example. We give partial answers to that problem and show how the cohomologies of X and $X_{\mathbb{Q}}$ can be very different in some situations.

10.1   Definitions and First Properties

Let $\varphi_X : \wedge V \xrightarrow{\simeq} A_{PL}(X)$ be the minimal Sullivan model of a connected space, X, and let L_X denote the homotopy Lie algebra of $\wedge V$. Recall that the adjoint map

$$\widetilde{\varphi_X} : X \to \langle \wedge V \rangle := X_{\mathbb{Q}}$$

is called the *Sullivan rationalization* of X.

Definition A path connected space, X, is a *Sullivan rational space* if the Sullivan rationalization $\widetilde{\varphi_X} : X \to X_{\mathbb{Q}}$ is a homotopy equivalence.

Remarks

1. The condition that $\widetilde{\varphi_X}$ is a homotopy equivalence implies that the minimal Sullivan model of X directly computes both $H(X_{\mathbb{Q}})$ and $\pi_*(X_{\mathbb{Q}})$.
2. Let Y be a Sullivan rational space, and let $f : X \to Y$ be a continuous map from a connected space. Since $f_{\mathbb{Q}} \circ \widetilde{\varphi_X} \sim \widetilde{\varphi_Y} \circ f$,

© The Author(s), under exclusive license to Springer Nature Switzerland AG 2026
Y. Félix, S. Halperin, *Lie Models for Spaces*, Frontiers in Mathematics,
https://doi.org/10.1007/978-3-032-15357-9_10

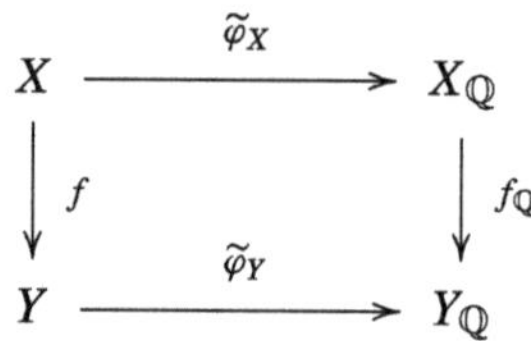

and $\widetilde{\varphi}_Y$ is a homotopy equivalence, it follows that $f \mapsto f_{\mathbb{Q}}$ yields an injection $[X, Y] \to [X_{\mathbb{Q}}, Y_{\mathbb{Q}}]$.

3. When X is Sullivan rational, the homotopy equivalence $X \xrightarrow{\simeq} X_{\mathbb{Q}}$ induces a homotopy equivalence $X_{\mathbb{Q}} \xrightarrow{\simeq} (X_{\mathbb{Q}})_{\mathbb{Q}}$. It follows that $X_{\mathbb{Q}}$ is also Sullivan rational.

Proposition 10.1 *Suppose* $\varphi : \wedge V \to A_{PL}(X)$ *is a morphism from a minimal Sullivan algebra. If X is Sullivan rational, then*

$$\varphi \text{ is a Sullivan model} \iff \pi_*(\widetilde{\varphi}) \text{ is an isomorphism.}$$

Proof If φ is a Sullivan model, then by definition $\pi_*(\widetilde{\varphi})$ is an isomorphism. On the other hand, let $\psi : \wedge W \xrightarrow{\simeq} A_{PL}(X)$ be a minimal Sullivan model. Lifting φ through ψ provides a homotopy commutative diagram

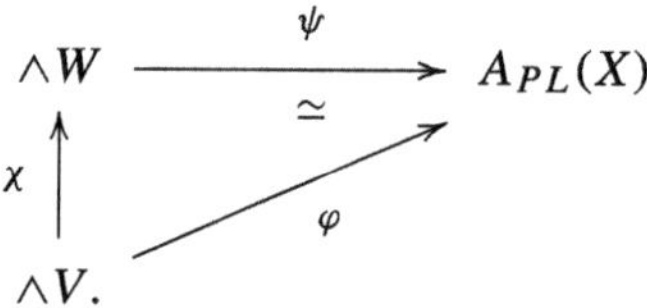

By adjonction, this gives the commutative diagram

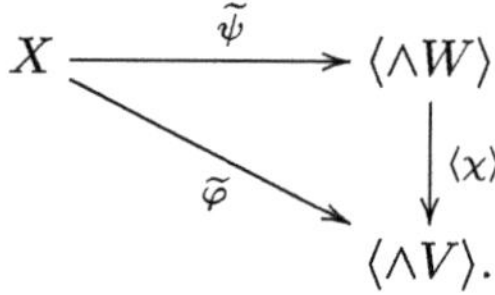

Since X is Sullivan rational, $\widetilde{\psi}$ is a homotopy equivalence. Therefore $\pi_*\chi$ is an isomorphism, and φ is a Sullivan model of X. $\qquad\qquad\square$

On the other hand, recall from Chap. 7 that adjoint to $\mathrm{id}_{\langle \wedge V \rangle}$ is the morphism

$$\iota_{\wedge V} : \wedge V \to A_{PL}\langle \wedge V \rangle.$$

Lemma 10.1 *Suppose $\varphi : \wedge V \to A_{PL}(X)$ is a Sullivan model of a connected space X. Then $\widetilde{\varphi}_X$ is a rational homotopy equivalence (i.e., induces an isomorphism in cohomology with rational coefficients) if and only if $\iota_{\wedge V}$ is a quasi-isomorphism.*

Proof According to Lemma 7.1, $\varphi = A_{PL}(\widetilde{\varphi}) \circ \iota_{\wedge V}$,

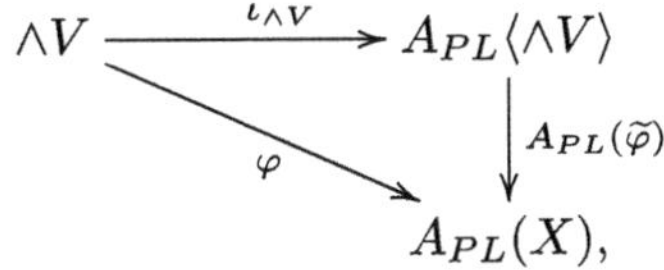

Since $H(\varphi)$ is an isomorphism, this establishes the Lemma. $\qquad\square$

Proposition 10.2 *For any minimal Sullivan algebra, $\wedge V$,*

$$\langle \wedge V \rangle \text{ is Sullivan rational} \iff \iota_{\wedge V} : \wedge V \to A_{PL}\langle \wedge V \rangle \text{ is a quasi-isomorphism.}$$

Moreover, if $\iota_{\wedge V}$ is a quasi-isomorphism, then:

(i) A basis of $\wedge V$ is at most countable.
(ii) $H(\wedge V)$ is a graded vector space of finite type.
(iii) $\pi_{\geq 2}\langle \wedge V \rangle$ is a rational vector space.
(iv) $H_1(\langle \wedge V \rangle; \mathbb{Z})$ is a finite dimensional rational vector space.

Remark In other words, $X_{\mathbb{Q}}$ is a Sullivan rational space if and only if $X \to X_{\mathbb{Q}}$ induces an isomorphism in homology with rational coefficients.

Proof Let $\wedge W \xrightarrow{\simeq} A_{PL}\langle \wedge V \rangle$ be the Sullivan minimal model of $\langle \wedge V \rangle$. By the lifting homotopy property, we deduce the following homotopy commutative diagram:

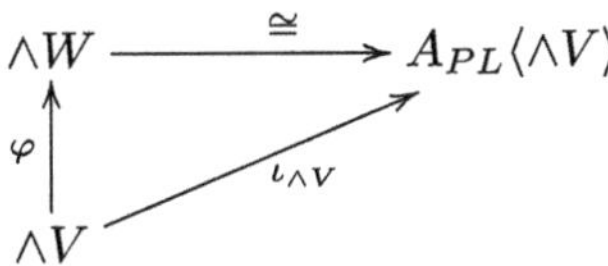

When $\iota_{\wedge V}$ is a quasi-isomorphism, the induced morphism $\varphi : \wedge V \to \wedge W$ is a quasi-isomorphism, and therefore $\langle \wedge V \rangle \to \langle \wedge W \rangle = \langle \wedge V \rangle_{\mathbb{Q}}$ is a homotopy equivalence.

In the reversed direction, if $\langle \wedge V \rangle$ is Sullivan rational, then the adjoint of the above diagram is a commutative diagram of the form

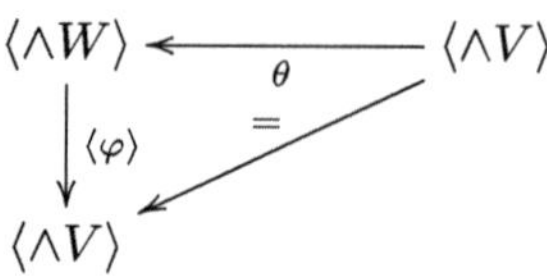

Since $\langle \wedge V \rangle$ is Sullivan rational, θ is a homotopy equivalence, and therefore $\langle \varphi \rangle$ is a homotopy equivalence. It follows that $\iota_{\wedge V}$ is a quasi-isomorphism.

For the rest of the proof we suppose $\iota_{\wedge V}$ is a quasi-isomorphism.

(i) It follows from (8.3) that the surjection $V^{\vee} \to (V \cap \ker d)^{\vee}$ factors as

$$V^{\vee} \xrightarrow{\ hur\ } H_*(\langle \wedge V \rangle; \mathbb{Q}) \longrightarrow (V \cap \ker d)^{\vee}.$$

Thus $H_*(\langle \wedge V \rangle; \mathbb{Q}) \to (V \cap \ker d)^{\vee}$ is surjective. Dualizing provides injections

$$(V^n \cap \ker d)^{\vee\vee} \to H^n \langle \wedge V \rangle \cong H^n(\wedge V).$$

Now assume by induction that $V^{<n}$ has an at most countable basis; then a cardinality argument implies that $V^n \cap \ker d$ must be finite dimensional.

Next observe that $V^n = \cup_{p \geq 0} V_p^n$, where

$$V_0^n = V^n \cap d^{-1}(\wedge V^{<n}) \quad \text{and} \quad V_{p+1}^n = V^n \cap d^{-1}(\wedge V^{<n} \otimes \wedge V_p^n).$$

Since $\dim V^n \cap \ker d < \infty$ and $\wedge V^{<n}$ has an at most countable basis, it follows by induction on p that V^n has an at most countable basis.

(ii) Since $H^n(\wedge V) \cong H^n(\langle \wedge V \rangle)$, which is the dual of $H_n\langle \wedge V \rangle$, and since $H^n(\wedge V)$ has an at most countable basis, it follows that each $\dim H^n(\wedge V) < \infty$.

(iii) This is Proposition 8.2(i).

(iv) The restriction of d to V^1 is the quadratic differential $d_1 : V^1 \to \wedge^2 V^1$. Since $\dim V^1 \cap \ker d < \infty$, we may apply [36, Chap 2] verbatim. In particular, Corollary 2.4 together with Theorem 2.4 of [36] provides isomorphisms

$$H_1(\langle \wedge V \rangle; \mathbb{Z}) = \frac{\pi_1 \langle \wedge V \rangle}{[\pi_1 \langle \wedge V \rangle, \pi_1 \langle \wedge V \rangle]} = G_L/[G_L, G_L] = L_0/[L_0, L_0],$$

where L is the homotopy Lie algebra of $\wedge V$ and $[\ ,\ ]$ denotes the normal subgroup or Lie ideal generated by commutators.

This identifies $H_1(\langle \wedge V \rangle; \mathbb{Z})$ as a rational vector space, whose dual is the finite dimensional vector space $H^1(\wedge V)$. $\qquad\square$

Example $X = (S^1 \vee S^2)_{\mathbb{Q}}.$

Lemma 10.2 *The cohomology $H^2(X)$ is uncountable, and the space X is not Sullivan rational.*

The Sullivan minimal model of $S^1 \vee S^2$ has the form $(\wedge V, d)$ where V has one generator t in degree 1, a family $(v_n)_{n \geq 0}$ of generators in degree 2 with $dv_0 = 0$, and for $n > 0$, $dv_n = tv_{n-1}$, and other generators in degrees >2.

The dual space $W = (V^2)^{\vee}$ can be identified with the space of series in one variable, $\mathbb{Q}[[t]]$: To $f \in W$ we associate the series $\sum f(v_n)t^n$. By [36, Theorem 4.1, page 126] the action of $\alpha \in \mathbb{Q} = \pi_1(\langle \wedge V^1 \rangle)$ on W is given by the multiplication by $e^{\alpha t}$

$$\alpha \cdot f = e^{\alpha t} f(t).$$

We denote by $[f]$ the vector space generated by the orbit of $[f]$ along this action

$$[f] = \left\{ \sum_{j=1}^{n} \beta_j e^{\alpha_j t} f(t) , \alpha_j, \beta_j \in \mathbb{Q}, n \geq 1 \right\}.$$

Two series f and f' are called equivalent if there is a rational number α with $f = e^{\alpha t} f'$. Clearly $f \sim f'$ if and only if $[f] = [f']$.

Recall that by construction $W = \pi_2(X) = \pi_2(\widetilde{X})$, where $\widetilde{X}$ is the universal cover of X. Then, $W \cong H_2(\widetilde{X})$ is an isomorphism of $\pi_1(X)$-module, and $W^{\vee} \cong H^2(\widetilde{X})$ with the induced $\pi_1(X)$-action.

The space W is uncountable. It follows that there is an uncountable set of nonequivalent series. Let f_0 be one of those series, and let Z be the vector space generated by the other nonequivalent series. Since the series $e^{\alpha t}$, $\alpha \in \mathbb{Q}$, are linearly independent, the elements $e^{\alpha t} f(t)$, $\alpha \in \mathbb{Q}$, form a basis of $[f_0]$.

We construct a linear map $\varphi : W \to \mathbb{Q}$, by putting $\varphi(Z) = 0$ and $\varphi(e^{\alpha t} f_0) = 1$. By construction φ is invariant along the action of $\pi_1(S^1_{\mathbb{Q}})$, $\varphi(\alpha \cdot f_0) = \varphi(f_0)$. The same is true for any class $[f]$. It follows that $H^2(\widetilde{X})$ contains an uncountable set of invariant elements.

Now we consider the Serre spectral sequence associated with the homotopy fibration $\widetilde{X} \to X \to S^1$. Since $E_2^{0,2}$ is the subspace of $H^2(\widetilde{X})$ generated by the invariant elements, $E_2^{0,2}$ is an uncountable set. Since $E_2^{2,1} = E_2^{3,0} = 0$, the spectral sequence collapses, and $E_2^{0,2} \cong H^2(X)$.

Now it follows that the map $\iota_{\wedge V}$ cannot be an isomorphism, and so by Proposition 10.2, $X_{\mathbb{Q}}$ is not a rational Sullivan space. $\square$

10.2 Simply Connected Rational Spaces

Fix a minimal Sullivan algebra, $\wedge V$, satisfying $V^1 = 0$. Recall then from [36, Theorem 1.5, page 42] that in this case $\iota_{\wedge V}$ is a quasi-isomorphism if and only if V is of finite type. We deduce the following lemma:

Lemma 10.3 *The following conditions are equivalent:*

(i) $\langle \wedge V \rangle$ *is Sullivan rational.*
(ii) Each $\dim V^k < \infty$.
(iii) Each $\dim H^k(\wedge V) < \infty$.
(iv) Each $\pi_k \langle \wedge V \rangle$ *is a finite dimensional rational vector space.*

Proposition 10.3 *Suppose* $\varphi : \wedge V \xrightarrow{\simeq} A_{PL}(X)$ *identifies* $\wedge V$ *as the minimal Sullivan model of a simply connected space, X. Then the following conditions are equivalent:*

(i) X is Sullivan rational.
(ii) Each $\pi_k(X)$ *is a finite dimensional rational vector space.*
(iii) Each $H_k(X; \mathbb{Z})$, $k \geq 1$, *is a finite dimensional rational vector space.*

Proof (i) $\iff$ (ii). By definition (i) implies that $\widetilde{\varphi} : X \to \langle \wedge V \rangle$ is a homotopy equivalence. In particular, $\pi_k(\widetilde{\varphi}) : \pi_k(X) \xrightarrow{\cong} (V^k)^\vee$. Since (Proposition 10.2) $\langle \wedge V \rangle$ is Sullivan rational, (ii) follows from Lemma 10.3(ii).

(ii) $\iff$ (iii). This follows from the Serre spectral sequence arising from the cellular filtration of X.

(ii) and (iii) $\Rightarrow$ (i). Since $H(\varphi)$ is an isomorphism, it follows from (iii) that $H(\wedge V)$ is a graded vector space of finite type. Thus $\langle \wedge V \rangle$ is Sullivan rational (Lemma 10.3) and so by Proposition 10.2, $\iota_{\wedge V}$ is a quasi-isomorphism. Since (7.2) $\varphi = A_{PL}(\widetilde{\varphi}) \circ \iota_{\wedge V}$,

$$H(\widetilde{\varphi}) : H(\langle \wedge V \rangle) \xrightarrow{\cong} H(X).$$

Moreover, since $\langle \wedge V \rangle$ is Sullivan rational, the implication (i) $\Rightarrow$ (iii) shows that each $H_k(\langle \wedge V \rangle; \mathbb{Z})$, $k \geq 1$, is also a finite dimensional rational vector space. Since $H_k(\widetilde{\varphi}; \mathbb{Q})$ is an isomorphism so is $H_k(\widetilde{\varphi}; \mathbb{Z})$ and since $\langle \wedge V \rangle$ and X are simply connected, this implies that $\widetilde{\varphi}$ is a homotopy equivalence. $\qquad\square$

Proposition 10.3 has been partially generalized by J. Zhou [84].

Proposition 10.4 *Let* $(\wedge V, d)$ *be a minimal Sullivan algebra. If* $\langle \wedge V \rangle$ *is a Sullivan rational space and* $\dim V^1 < \infty$, *then* $H(\wedge V, d)$ *is a finite type graded vector space.*

Proof We suppose that $H(\wedge V, d)$ is not of finite type and arrive to a contradiction. Let thus $n > 1$ be the smallest integer such that $\dim H^n(\wedge V, d) = \infty$. Since by hypothesis $H^n(\wedge V) \cong H^n(\langle \wedge V \rangle) = (H_n(\langle \wedge V \rangle))^\vee$, $H^n(\wedge V)$ is uncountable. We denote its cardinality by $\aleph_0$.

Since V^1 and $H^{<n}(\wedge V)$ are of finite type, $V^{<n}$ is of finite type. We decompose V^n as usual as the union $V^n = \cup_{r \geq 0} V^n(r)$ with $V^n(0) = V^n \cap \ker d$, and for $r > 0$,

$$V^n(r) = \{x \in V^n \text{ such that } dx \in (\wedge V^{<n} \otimes \wedge V^n(r-1))\}.$$

Then the cardinality of $V^n(0)$ and of its image in $H^n(\wedge V)$ is $\aleph_0$.

Denote by $\xi : \wedge^{\geq 1} V \to V$ the natural projection. It follows that the image of the projection $H^n(\wedge V) \to V^n$ has also cardinality $\aleph_0$. Now from Proposition 8.3, we have the commutative diagram

$$
\begin{array}{ccc}
\pi_n\langle \wedge V \rangle & \xrightarrow[\cong]{\iota_n} & (V^n)^\vee \\
\downarrow{\scriptstyle hur} & & \downarrow{\scriptstyle (\xi^\vee} \\
H_n\langle \wedge V \rangle & \xrightarrow{\gamma} (H^n\langle \wedge V \rangle)^\vee \xrightarrow{(H^n(i_{\wedge V}))^\vee} & H^n(\wedge V, d))^\vee,
\end{array}
$$

where γ is the injection of a space in its bidual. Now the cardinality of the image of the image of $\xi^\vee$ is $\aleph_1 > \aleph_0$. It follows that the cardinality of $H^n\langle \wedge V \rangle$ is $\geq \aleph_1$, which is a contradiction. $\qquad\qquad\square$

Finally, as an immediate consequence of [36, Theorem 5.4], we obtain the following proposition:

Proposition 10.5 *If $\wedge V$ is a Sullivan algebra and each $\dim V^l < \infty$, then $\iota_{\wedge V}$ is a quasi-isomorphism, and so $\langle \wedge V \rangle$ is Sullivan rational.*

Conjecture If $\wedge V$ is a Sullivan algebra, then
$\langle \wedge V \rangle$ is Sullivan rational $\iff$ each $\dim V^\ell < \infty$.

10.3   Discrete Groups

Let

$$\varphi : \wedge V \to A_{PL}(BG)$$

be the minimal Sullivan model of the classifying space of a discrete group, G. Thus BG is Sullivan rational if and only if

$$\widetilde{\varphi} : BG \to \langle \wedge V \rangle$$

is a homotopy equivalence.

Proposition 10.6 *If BG is Sullivan rational. then $V = V^1$, $\dim G/[G, G] \otimes \mathbb{Q} < \infty$, and G is Malcev complete.*

Proof Let G_L be the fundamental group associated with the homotopy Lie algebra of $\wedge V$. Then suppose first that BG is Sullivan rational. In this case $\pi_*(\widetilde{\varphi})$ is an isomorphism, and it follows that

$$\pi_1(\widetilde{\varphi}) : G \xrightarrow{\cong} \pi_1\langle \wedge V \rangle = G_L \quad \text{and} \quad 0 = \pi_{\geq 2}\langle \wedge V \rangle = (V^{\geq 2})^\vee.$$

In particular $V = V^1$.

Moreover, since BG is Sullivan rational, it follows from Proposition 10.3 that $G/[G, G] = H_1(BG, \mathbb{Z})$ is a finite dimensional rational vector space. Therefore G_L is Malcev complete [36, Theorem 7.5], and since $G \cong G_L$, so is G. $\qquad\square$

10.4   Fibrations

Fix a fibration $B \leftarrow X \leftarrow F$ of path connected spaces. Then there is a commutative diagram of cdga morphisms [36, §3.7]

$$
\begin{array}{ccccc}
\wedge W & \longrightarrow & \wedge W \otimes \wedge Z & \longrightarrow & \wedge Z \\
{\scriptstyle \varphi_B}\downarrow{\scriptstyle \simeq} & & {\scriptstyle \varphi_X}\downarrow{\scriptstyle \simeq} & & {\scriptstyle \varphi_F}\downarrow \\
A_{PL}(B) & \longrightarrow & A_{PL}(X) & \longrightarrow & A_{PL}(F)
\end{array}
\tag{10.1}
$$

in which φ_B is a minimal Sullivan model and $\wedge W \otimes \wedge Z$ is a minimal Λ-extension of $\wedge W$.

Proposition 10.7 *Suppose $H(F)$ is a graded vector space of finite type. Then:*

(i) The holonomy representation of $\pi_1(B)$ in $H(F)$ is locally nilpotent if and only if $H(\varphi_F)$ is an isomorphism.

(ii) If $H(\varphi_F)$ is an isomorphism and if any two of B, X, F are Sullivan rational, then the third space is also Sullivan rational.

Proof

(i) According to [36, Theorem 5.1], if the holonomy representation of $\pi_1(B)$ is locally nilpotent, then $H(\varphi_F)$ is an isomorphism.

In the reverse direction, suppose $H(\varphi_F)$ is an isomorphism. Then the adjoint map $\widetilde{\varphi_B}$ induces a morphism [36, Cor. 2.7]

$$\pi_1(\widetilde{\varphi_B}) : \pi_1(B) \to \pi_1\langle\wedge W\rangle = G_L,$$

where L is the homotopy Lie algebra of $\wedge W$. Moreover, $H(\varphi_F)$ is equivariant with respect to the holonomy representations of $\pi_1(B)$ in $H(F)$ and of G_L in each $H^k(\wedge Z)$ [36, Theorem 4.2]. Since the holonomy representation of G_L is locally nilpotent, while the holonomy representation of $\pi_1(B)$ in each $H^k(F)$ is the dual of a representation in $H_k(F; \mathbb{Q})$ [36, Proposition 4.4], it follows that both representations are nilpotent.

(ii) Applying the adjoint construction to the diagram (10.1) yields the commutative diagram

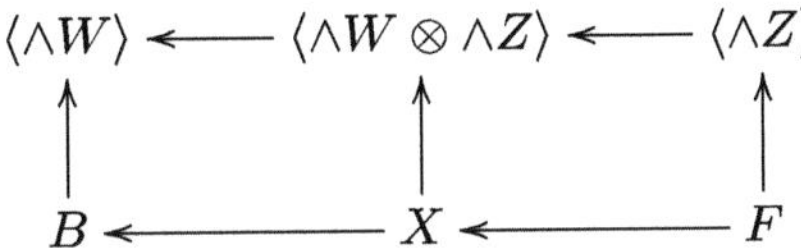

which in particular is a map of fibrations [36, Proposition 17.9]. Thus if any two of the vertical maps are a homotopy equivalence, so is the third.

$\square$

Remark When B is nilpotent and $H(F)$ is a finite type vector space, then $H(F)$ contains a maximal unipotent $\pi_1(B)$-module, U, called its fitting module [51]. In that case φ_F induces an isomorphism $H^*(\wedge Z) \cong U$ [32, Theorem 4].

Now consider more generally a fibration $\xi : F \to X \to B$ with B path connected and where $H(F)$ is a finite type graded vector space. For each $\alpha \in \pi_1(B)$ denote by U_α the fitting module of $H(F)$ for the action of α.

Proposition 10.8 *The image of* $\varphi_F : H(\wedge Z, d) \to H(F)$ *is contained in* $\cap_\alpha U_\alpha$.

Proof Denote by $F \to X_\alpha \to S^1$ the fibration induced by ξ along a representative of $\alpha : S^1 \to B$. The commutativity of the diagram

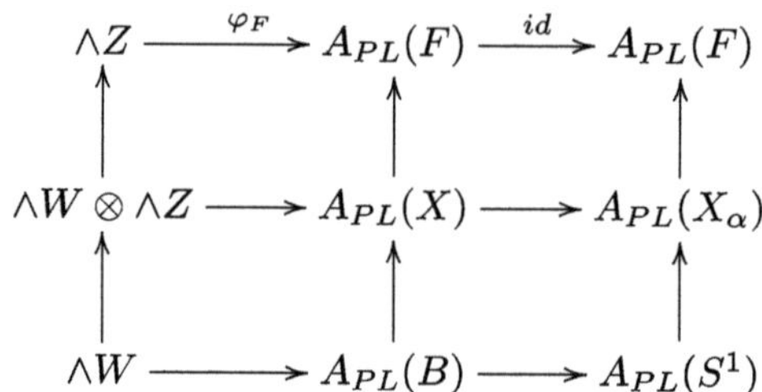

implies that the image of φ_F is contained in U_α. This is true for all α and that proves the result. $\qquad\square$

Part III

Enriched dgl's and Their Homotopy Theory

Enriched dgl's and Semi-quadratic Sullivan Algebras

11

Enriched dgl's are differential graded Lie algebras (L, ∂) that are inverse limits of finite dimensional nilpotent differential graded Lie algebras. We extend the constructions established before for enriched Lie algebras. In particular we construct an equivalence of categories between enriched dgl's and Sullivan algebras of the form $(\wedge V, d)$ where $d = d_0 + d_1$, $d_0 : V \to V$, and $d_1 : V \to \wedge^2 V$. The first properties of enriched dgl's are then described.

11.1 The Correspondence

Recall from the Introduction that a *differential graded Lie algebra* (dgl), (L, ∂), is a graded Lie algebra $L = \{L_k\}$ together with a differential $\partial : L_k \to L_{k-1}$ which is also a derivation:

$$\partial^2 = 0 \quad \text{and} \quad \partial[x, y] = [\partial x, y] + (-1)^{deg\,x}[x, \partial y]. \tag{11.1}$$

Now we introduce the subcategory of *enriched dgl's*, generalizing the enriched Lie algebras of Part I.

Definitions

(1) An *enriched dgl* is a dgl, (L, ∂), together with an inverse system of surjective morphisms $\rho_\alpha : (L, \partial) \to (L_\alpha, \partial_\alpha)$ such that:

- $L = L_{\geq 0}$.
- Each $(L_\alpha, \partial_\alpha)$ is a finite dimensional nilpotent dgl.

Y. Félix, S. Halperin, *Lie Models for Spaces*, Frontiers in Mathematics,
https://doi.org/10.1007/978-3-032-15357-9_11

- $L \xrightarrow{\cong} \varprojlim_\alpha L_\alpha.$

In particular, *L is an enriched Lie algebra.*
(2) An enriched dgl, (L, ∂) is *minimal* if $\partial : L \to L^{(2)}$.
(3) A *morphism of enriched dgl's* is a morphism of enriched Lie algebras which commutes with the differentials. We denote the set of morphisms $(L, \partial) \to (L', \partial')$ by $\mathrm{dgl}((L, \partial), (L', \partial'))$.
(4) A *dgl ideal* in an enriched dgl, L, is a closed differential ideal; in which case (Lemma 1.2) L/I is an enriched dgl.

For simplicity we will frequently denote an enriched dgl simply by (L, ∂).

In this part we shall frequently rely on the properties of enriched vector spaces and on the duality between enriched Lie algebras and quadratic Sullivan algebras. In particular, the 1-1 correspondence in Chap. 4 between enriched Lie algebras and quadratic Sullivan algebras extends as follows to a 1-1 correspondence between enriched dgl's and semi-quadratic Sullivan algebras, with the following definition:

Definitions

(i) A *semi-quadratic Sullivan algebra* is a Sullivan algebra of the form $(\wedge V, d_0 + d_1)$, where $d_0 : V \to V$ and $d_1 : V \to \wedge^2 V$.
(ii) A cdga morphism $\varphi : (\wedge V, d_0 + d_1) \to (\wedge W, d_0 + d_1)$ between semi-quadratic Sullivan algebras is *bihomogeneous* if $\varphi : V \to W$.

Let (L, ∂) be an enriched dgl with enriched structure provided by surjections $\rho_\alpha : (L, \partial) \to (L_\alpha, \partial_\alpha)$. In particular, L is an enriched Lie algebra with quadratic Sullivan model

$$(\wedge V, d_1) = \varinjlim_\alpha (\wedge V_\alpha, d_{1,\alpha}),$$

where $(\wedge V_\alpha, d_{1,\alpha}) = C^*(L_\alpha)$.
We define the generalized cochain construction $\widehat{C}(L, \partial)$:

$$\widehat{C}(L, \partial) = (\wedge V, d_0 + d_1) := \varinjlim_\alpha C^*(L_\alpha, \partial_\alpha) = \varinjlim_\alpha (\wedge V_\alpha, d_{0\alpha} + d_{1\alpha}).$$

Here ∂_α is the dual (up to suspension) of a unique differential $d_{0,\alpha}$ in V_α. It follows from Lemma 11.1 below that $(\wedge V, d_0 + d_1)$ is a semi-quadratic Sullivan algebra.

In particular, (V, d_0) is the complex of morphisms of enriched vector spaces from (L, ∂) to $(\mathbb{Q}, 0)$.

This gives the natural linear isomorphisms

$$s(L, \partial) = \varprojlim_{\alpha} s(L_\alpha, \partial_\alpha) = [\varinjlim_{\alpha}(V_\alpha, d_{0,\alpha})]^\vee = (V, d_0)^\vee$$

and

$$sH(L, \partial) = H(V, d_0)^\vee$$

$$\tag{11.2}$$

Theorem 11.1

(i) *The cochain construction* $\widehat{C}$ *induces a bijection between enriched dgl's and semi-quadratic Sullivan algebras.*

(ii) *This correspondence identifies the morphisms* $\ell : (L, \partial) \rightarrow (L', \partial')$ *with the bihomogeneous morphisms* $\varphi : (\wedge V, d_0 + d_1) \leftarrow (\wedge V', d_0' + d_1')$ *via*

$$s\ell = \varphi^\vee : (V')^\vee \rightarrow ((V')^\vee.$$

In particular ℓ *is a quasi-isomorphism if and only if* φ *is a quasi-isomorphism.*

(iii) (L, ∂) *is minimal if and only if* $d_0(V \cap \ker d_1) = 0.$

(iv) *If* $I \subset L$ *is a dgl ideal in an enriched dgl,* L, *then the morphisms* $\wedge V_{L/I} \rightarrow \wedge V_L \rightarrow \wedge V_I$ *decompose* $\wedge V_L$ *as a Sullivan extension*

$$\wedge V_L = \wedge V_{L/I} \otimes \wedge V_I.$$

Conversely, if an inclusion $\wedge W \rightarrow \wedge V_L$ *of semi-quadratic Sullivan algebras decomposes* $\wedge V_L$ *as a Sullivan extension* $\wedge W \otimes \wedge Z$, *then the enriched dgl corresponding to* $\wedge Z$ *is a dgl ideal* $I \subset L$ *and* $\wedge W = \wedge V_{L/I}.$

Proof

(i) This is immediate.

(ii) The correspondence associates with a morphism ℓ the map $\varphi : V \leftarrow V'$, via the equation

$$< \varphi v', sx >=< v', s\ell(x) > .$$

In this case $\varphi \circ d_0 = d_0' \circ \varphi$ and $\varphi \circ d_1 = d_1' \circ \varphi$. The bijection is a straightforward consequence of the definitions and Propositions 2.1 and 4.1.

(iii) This follows because (Lemma 4.1) the isomorphism $sL \cong V^\vee$ induces an isomorphism $s(L/L^{(2)}) \cong (V \cap \ker d_1)^\vee.$

(iv) This is immediate from the definitions and Sect. 4.1. $\qquad\square$

Definition In the correspondence $(L, \partial) \leftrightarrow (\wedge V_L, d_0 + d_1)$ of Proposition 11.1, (L, ∂) is the *enriched dgl model* of $(\wedge V_L, d_0 + d_1)$, and $(\wedge V_L, d_0 + d_1)$ is the *semi-quadratic Sullivan model* of $(L, \partial).$

Remark In working with enriched dgl's and their semi-quadratic models, it is important to note that:

(1) Not every morphism of semi-quadratic Sullivan algebras is bihomogeneous.
(2) The semi-quadratic Sullivan model of a minimal enriched dgl is a Sullivan algebra but is not minimal unless $d_0 = 0$.

11.2   The Homology of an Enriched dgl

Suppose (L, ∂) is an enriched dgl. Then let

$$\varphi : (\wedge V, d_0 + d_1) \xrightarrow{\; \cong \;} (\wedge W, d)$$

be a surjective quasi-isomorphism from its associated semi-quadratic Sullivan algebra to its minimal Sullivan model.

The filtration of $\wedge V$ by the ideals $\wedge^{\geq k} V$ induces a spectral sequence with $(E_1, \overline{d_1}) = (\wedge H(V, d_0), \overline{d_1})$. Thus $(\wedge H(V, d_0), \overline{d_1})$ is a cdga.

Lemma 11.1 *With the hypotheses and notation above:*

 (i) A semi-quadratic algebra $(\wedge V, d_0 + d_1)$ is a Sullivan algebra if and only if $(\wedge V, d_1)$ is a Sullivan algebra.
 (ii) If $(\wedge V, D)$ is a Sullivan algebra, let $(\wedge W, D_W)$ be its minimal Sullivan model, $D_W = D_{W,1} + D_{W,2} + \ldots$ with $D_{W,n}(W) \subset \wedge^{n+1} W$. Then $(\wedge H(V, d_0), \overline{d_1})$ is isomorphic to $(\wedge W, D_{W,1})$.

Proof

(i) If $(\wedge V, d_0 + d_1)$ is a Sullivan algebra, then $V = \varinjlim_n V_n$ with $d_0 + d_1 : V_{n+1} \to \wedge V_n$ and $D(V_1) = 0$. It follows that $d_1 : V_{n+1} \to \wedge V_n$, and so $(\wedge V, d_1)$ is a Sullivan algebra.

   In the reverse direction suppose $(\wedge V, d_1)$ is a Sullivan algebra. Then V is the direct limit of finite dimensional subspaces P such that $\wedge P$ is preserved by d_1. Therefore each $\wedge (P \oplus d_0 P)$ is preserved by $d_0 + d_1$.

(ii) Decompose V as $W \oplus V' \oplus d_0 V'$ with $W \subset \ker d_0$ and $d_0 : V' \xrightarrow{\; \cong \;} d_0 V$. The inclusions $W, V', DV' \to \wedge V$ then extend to a morphism of graded algebras

$$\varphi : \wedge W \otimes \wedge V' \otimes \wedge DV' \xrightarrow{\; \cong \;} \wedge V.$$

By construction, φ preserves filtration by wedge degree. Moreover, the induced map of the associated bigraded algebras is the identity isomorphism

$$\wedge W \otimes \wedge V' \otimes \wedge d_0 V' \xrightarrow{\;\cong\;} \wedge V.$$

It follows that φ is an isomorphism and so a Sullivan algebra $(\wedge W \otimes \wedge V' \otimes \wedge DV', \delta)$ is defined by $\varphi \circ \delta = D \circ \varphi$. In particular $\delta = D : V' \xrightarrow{\cong} DV'$. Since $(\wedge V, D)$ is a Sullivan algebra, so is $(\wedge W \otimes \wedge V' \otimes \wedge DV', \delta)$. Now a sequence

$$(\wedge V, D) \xrightarrow[\varphi^{-1}]{\;\cong\;} (\wedge W \otimes \wedge V' \otimes \wedge DV', \delta) \xrightarrow{\;\simeq\;} (\wedge W, D_W)$$

is defined by division by V' and DV'. By construction, $\delta : \wedge W \to \wedge^{\geq 2}(W \oplus V' \oplus DV')$, and so the sequence above identifies $(\wedge W, D_W)$ as the minimal Sullivan model of $(\wedge V, D)$.

Finally, since φ preserves filtrations and induces an isomorphism of the associated bigraded algebras, it induces an isomorphism of the E_1-terms of the associated spectral sequences. This has the form

$$(\wedge W, D_{W,1}) \xleftarrow{\;\cong\;} (\wedge H(V, d_0), \overline{d}_1),$$

which identifies $(H(\wedge V, d_0), \overline{d}_1)$ as the quadratic Sullivan algebra associated with L_W.

$\square$

Proposition 11.1 *Suppose (L, ∂) is an enriched dgl. Then let*

$$\varphi : (\wedge V_L, d_0 + d_1) \xrightarrow{\;\simeq\;} (\wedge W, d)$$

be a surjective quasi-isomorphism from its associated semi-quadratic Sullivan algebra to its minimal Sullivan model.

(i) $H(L)$ is the homotopy Lie algebra of the quadratic Sullivan algebra $(\wedge H(V_L, d_0), \overline{d}_1)$.

(ii) φ induces a natural isomorphism $H(L) \xleftarrow{\;\cong\;} L_W$ of Lie algebras.

(iii) $H(L)$ is a natural enriched Lie algebra. If $(L, \partial) = \varprojlim_\beta (L_\beta, \partial_\beta)$, then $H(L) = \varprojlim_\beta H(L_\beta)$.

Proof

(i) A standard limit argument reduces this to the case that L is finite dimensional and nilpotent. In this case there is a short exact sequence $(\mathbb{Q}z, 0) \to L \to L/z$, and so (i) follows by induction on $\dim L$.
(ii) This is Lemma 11.1 (ii).
(iii) It follows that $(V_L, d_0) = \varinjlim_\beta (V_\beta, d_0)$ and therefore that $H(V_L, d_0) = \varinjlim_\beta H(V_\beta, d_0)$. This gives $H(L) = \varprojlim_\beta H(L_\beta)$.

$\square$

11.3   Homotopy of Morphisms of Semi-quadratic Sullivan Algebras

If $(\wedge V, d_0 + d_1)$ is a semi-quadratic Sullivan algebra, then it is immediate that $\wedge V(n) :=$ $\wedge V^{\leq n} \otimes \wedge d_0 V^n$ is preserved by both d_0 and d_1. Thus

$$(\wedge V, d_0 + d_1) = \varinjlim_n (\wedge V(n), d_0 + d_1).$$

Proposition 11.2

(i) *A bihomogeneous morphism $\varphi : (\wedge V, d_0 + d_1) \to (\wedge W, d_0 + d_1)$ of semi-quadratic Sullivan algebras restricts to morphisms*

$$\varphi(n) : (\wedge V(n), d_0 + d_1) \to (\wedge W(n), d_0 + d_1), \qquad n \geq 0.$$

(ii) *If $\psi \sim \varphi : (\wedge V, d_0 + d_1) \to (\wedge W, d_0 + d_1)$, then*

$$\varphi(n) \sim \psi(n), \qquad n \geq 0.$$

(iii) *A bihomogeneous morphism $\varphi : (\wedge V, d_0 + d_1) \to (\wedge W, d_0 + d_1)$ is a quasi-isomorphism if and only if each $\varphi(n)$ is a quasi-isomorphism.*

Proof

(i) Since φ preserves degrees, it maps $\wedge V^{\leq n}$ to $\wedge W^{\leq n}$. But if $v \in V^n$, then $\varphi(d_0 v) \in W(n)$, and so $\varphi : V(n) \to W(n)$.
(ii) This follows in the same way as (i).
(iii) When φ is a quasi-isomorphism of Sullivan algebras, it has a homotopy inverse $\widehat{\varphi} :$ $\wedge W \to \wedge V$. It follows from (i) and (ii) that $\widehat{\varphi}$ and φ restrict to homotopy inverses between each $V(n)$ and $W(n)$.

$\square$

Proposition 11.3 *Suppose $\varphi : (\wedge V, d_0 + d_1) \leftarrow (\wedge V', d_0' + d_1')$ is the bihomogeneous morphism associated with a morphism $\ell : (L, \partial) \to (L', \partial')$ of enriched dgl's. Then:*

(i) $H(\varphi, d_0) : (\wedge H(V, d_0), \overline{d_1}) \leftarrow (\wedge H(V', d_0'), \overline{d_1'})$ is the bihomogeneous morphism corresponding to $H(\ell) : (H(L), 0) \to (H(L'), 0)$.

(ii) $H(\ell)$ is an isomorphism $\Leftrightarrow$ $H(\varphi, d_0)$ is an isomorphism $\Leftrightarrow$ $H(\varphi, d_0 + d_1)$ is an isomorphism.

Proof

(i) It follows from (11.2) that $H(\varphi, d_0)^{\vee}$ is the suspension of $H(\ell) : L_V \to L_{V'}$, where L_V and $L_{V'}$ are the homotopy Lie algebras of $\wedge H(V, d_0)$ and $\wedge H(V', d_0')$.

(ii) The first equivalence follows immediately from (i). On the other hand convert $\varphi : (\wedge V, d_0 + d_1) \leftarrow (\wedge V', d_0' + d_1')$ up to equivalence to a morphism $\chi : (\wedge W, D) \leftarrow (\wedge W', D')$. Since these are minimal Sullivan algebras, χ is an isomorphism $\Leftrightarrow H(\chi)$ is an isomorphism, and this gives the second equivalence. $\qquad\square$

Proposition 11.4 *Suppose $\wedge V$ is a semi-quadratic Sullivan algebra associated with a profree dgl, and $\wedge V'$ is another semi-quadratic Sullivan algebra. Then any morphism $\varphi : \wedge V' \to \wedge V$ is homotopic to a bihomogeneous morphism.*

Proof It is sufficient to consider the case that $\wedge V' = \wedge W \otimes \wedge Z$ in which $d : Z \to \wedge W$ and $\varphi|_{\wedge W}$ is bihomogeneous and then to construct

$$\psi : \wedge W \otimes \wedge Z \to \wedge V \otimes \wedge(t, dt)$$

so that $\psi|_{\wedge W} = \varphi$, $\varepsilon_0 \circ \psi = \varphi$, and $\varepsilon_1 \circ \varphi : Z \to V$.

Fix a basis of Z, and let z be an element of that basis. Write $\varphi(z) = \sum_{n \leq n_0} u_n$ with $u_n \in \wedge^n V$. Since $\wedge V'$ is semi-quadratic and $\varphi|_{\wedge W}$ is bihomogeneous, it follows that $\varphi(dz)$ is a cycle. Since $d\varphi z = \varphi dz$, if $n_0 \geq 2$, u_{n_0} is a d_1-cycle in $\wedge^{\leq 2} V$ and so a d_1-boundary, $u_{n_0} = d_1 a$. We consider then the homotopy

$$\Psi : \wedge W \otimes \wedge Z \to \wedge V \otimes \wedge(t, dt)$$

defined by $\Psi(w) = \varphi(w)$ and $\Psi(z) = \varphi z - d(ta)$. Since $\varphi \sim \varepsilon_1 \Psi$, we can suppose by induction that $n_0 = 1$. This implies the result. $\qquad\square$

Remark The hypothesis on $\wedge V'$ is necessary. For instance, the morphism $\varphi : (\wedge u, 0) \to (\wedge(x, y, z), 0)$ defined by $\varphi(u) = xyz$, $\deg u = 3$, $\deg x = 1$, $\deg y = 1$, $\deg z = 1$, is a morphism between semi-quadratic Sullivan algebras that is not homotopic to a bihomogeneous morphism.

11.4   Free Products of Enriched dgl's

The differentials in enriched dgl's (L', ∂') and (L'', ∂'') extend uniquely to a differential, ∂, in the free product $L' \mathbin{\widehat{\amalg}} L''$ (defined in Sect. 6.6). Then $(L = L' \mathbin{\widehat{\amalg}} L'', \partial)$ is an enriched dgl: the *free product* of (L', ∂') and (L'', ∂'').

Proposition 11.5

(i) *Any two morphisms,* ι', ι'' : $(L', \partial'), (L'', \partial'') \to (E, \partial_E)$, *extend uniquely to a morphism*

$$(L' \mathbin{\widehat{\amalg}} L'', \partial) \to (E, \partial_E).$$

(ii) *If* $(\wedge W, d_0 + d_1)$ *and* $(\wedge Q, d_0 + d_1)$ *and* $(\wedge V, d_0 + d_1)$ *are the semi-quadratic models of* (L', ∂'), (L'', ∂''), *and* $(L' \mathbin{\widehat{\amalg}} Lv'', \partial)$, *then there is a bihomogeneous quasi-isomorphism*

$$(\wedge V, d_0 + d_1) \xrightarrow{\ \simeq\ } (\wedge W, d_0 + d_1) \times_{\mathbb{Q}} (\wedge Q, d_0 + d_1).$$

Proof

(i) This follows from Lemma 6.3 since by construction the morphism $L' \mathbin{\widehat{\amalg}} L'' \to E$ is compatible with the differentials.

(ii) The proof of Proposition 8.7 provides a quasi-isomorphism

$$\varphi : (\wedge V, d_1) \to (\wedge W, d_1) \times_{\mathbb{Q}} (\wedge Q, d_1)$$

in which $(\wedge V, d_1)$ is a quadratic Sullivan algebra and $\varphi : V \to W \oplus Q$. Let L be the homotopy Lie algebra of $(\wedge V, d_1)$. Then the surjections $\wedge V \to \wedge W, \wedge Q$ induce an isomorphism (Proposition 8.8)

$$L' \mathbin{\widehat{\amalg}} L'' \xrightarrow{\ \cong\ } L,$$

of enriched Lie algebras.

In particular, ∂' and ∂'' extend to a differential, ∂, in L. This in turn defines a differential d_0 in $\wedge V$ for which $(\wedge V, d_0 + d_1)$ is a semi-quadratic Sullivan algebra and $\varphi \circ d_0 = d_0 \circ \varphi$. Filtering by degree $-$ wedge produces a morphism of spectral sequences converging from $H(\varphi, d_1)$ to $H(\varphi, d_0 + d_1)$. Thus

$$\varphi : (\wedge V, d_0 + d_1) \to (\wedge W, d_0 + d_1) \times_{\mathbb{Q}} (\wedge Q, d_0 + d_1)$$

is a quasi-isomorphism. $\square$

11.5   The Universal Enveloping Algebra of an Enriched dgl

The universal enveloping algebra, UL, of a differential graded Lie algebra, (L, ∂) is naturally a differential graded algebra (UL, ∂): The differential is the unique derivation extending the differential in L. As noted in [36, Theorem 21.7], the inclusion $L \to UL$ induces a Lie bracket preserving map $H(L) \to H(UL)$ which then extends to an isomorphism $UH(L) \overset{\cong}{\longrightarrow} H(UL)$ of graded algebras. The proof of that theorem also shows that this induces isomorphism

$$UH(L)/I_H^n \overset{\cong}{\longrightarrow} H(U(L/I^n)), \tag{11.3}$$

where I_H^n and I^n denote the nth powers of the respective augmentation ideals.

Now suppose surjections $\rho_\alpha : (L, \partial) \to (L_\alpha, \partial_\alpha)$ define the enriched structure of an enriched dgl (L, ∂). Then recall that the completion $\overline{UL}$ of UL is defined as

$$\overline{UL} = \varprojlim_{n,\alpha} U(L_\alpha/J_\alpha^n),$$

J_α^n denoting the nth power of the augmentation ideal. On the other hand, by Proposition 11.1, $H(L)$ is an enriched graded Lie algebra, and we can form the completion

$$\overline{UH(L)} = \varprojlim_{n,\alpha} UH(L_\alpha)/I_{\alpha,H}^n,$$

where $I_{\alpha,H}$ denotes the augmentation ideal of $H(L_\alpha)$.

The next proposition is a generalization of Theorem 21.7 in [33].

Proposition 11.6 *There is a natural isomorphism*

$$\overline{UH(L)} \overset{\cong}{\longrightarrow} H(\overline{UL}).$$

Proof This follows directly from (11.3) together with Proposition 1.1 via the isomorphisms

$$\overline{UH(L)} = \varprojlim_{n,\alpha} UH(L_\alpha)/I_{\alpha,H}^n = \varprojlim_{n,\alpha} H(U(L_\alpha/J_\alpha^n)) = H(\overline{UL}).$$

$\square$

Profree dgl's and Profree dgl Models **12**

A profree dgl is a dgl (L, ∂), where L is a profree Lie algebra. A profree dgl model of an enriched dgl (L, ∂) is a profree dgl $(\overline{\mathbb{L}}_V, \partial)$ quasi-isomorphic to (L, ∂). It is minimal if $\partial(V) \subset \mathbb{L}_V^{\geq 2}$. We prove that each enriched dgl admits a unique, up to isomorphism, minimal profree dgl model.

As in Part I, $\mathbb{L}_T$ denotes the free graded Lie algebra freely generated by an enriched vector space $T = \varprojlim_\alpha T_\alpha$. The profree Lie algebra $\overline{\mathbb{L}}_T$ is its completion for the projections $\mathbb{L}_T \to \mathbb{L}_{T_\alpha}/\mathbb{L}_{T_\alpha}^n$.

In this case the surjection

$$\rho : L \to L/L^{(2)}$$

is split: $L = T \oplus L^{(2)}$ and T is closed in L. Finally Theorem 6.1, together with Lemma 6.2, establishes the equivalence of the following three conditions on an enriched Lie algebra, L, and its quadratic model $(\wedge V, d_1)$:

$$
\left.
\begin{array}{ll}
(i) & L = \overline{\mathbb{L}}_T \text{ is profree.} \\
(ii) & H(\wedge V, d_1) = \mathbb{Q} \oplus (V \cap \ker d_1). \\
(iii) & H^{[2]}(\wedge V, d_1) = 0.
\end{array}
\right\}
\qquad (12.1)
$$

(Here $H(\wedge V, d_1) = \oplus_k H^{[k]}(\wedge V, d_1)$ is the decomposition induced by wedge degree.)

Now we pass to profree dgl's.

Y. Félix, S. Halperin, *Lie Models for Spaces*, Frontiers in Mathematics,
https://doi.org/10.1007/978-3-032-15357-9_12

Definition

(i) A *profree dgl* is an enriched dgl (L, ∂) in which $L \cong \overline{\mathbb{L}}_T$ is profree. In particular, division by $\overline{\mathbb{L}}_T^{(2)}$ is a surjective morphism of enriched vector spaces:

$$\rho : (\overline{\mathbb{L}}_T, \partial) \to (T, \overline{\partial}).$$

(ii) The profree dgl (L, ∂) is *minimal* if $\overline{\partial} = 0$.

(iii) A *(minimal) profree dgl model of an enriched dgl* (L, ∂) is a quasi-isomorphism

$$f : (\overline{\mathbb{L}}_T, \partial) \xrightarrow{\simeq} (L, \partial)$$

from a (minimal) profree dgl.

Proposition 12.1 *Suppose* $(\overline{\mathbb{L}}_T, \partial)$ *is a profree dgl with semi-quadratic Sullivan model* $(\wedge V, d_0 + d_1)$. *Let* $\xi : \wedge^{\geq 1} V \to V$ *denote the surjection with kernel* $\wedge^{\geq 2} V$. *Then:*

(i) *Any morphism* $(T, \overline{\partial}) \to (L, \partial)$ *of an enriched vector space into an enriched dgl extends uniquely to a morphism* $(\overline{\mathbb{L}}_T, \partial) \to (L, \partial)$.

(ii) $H(V \cap \ker d_1, d_0) \xrightarrow{\cong} H^{\geq 1}(\wedge V, d_0 + d_1)$.

(iii) *The isomorphism (21),* $s(\overline{\mathbb{L}}_T, \partial) \xrightarrow{\cong} (V, d_0)^\vee$, *extends to the commutative diagram*

$$
\begin{array}{ccc}
s(\overline{\mathbb{L}}_T, \partial) & \xrightarrow{\cong} & (V, d_0)^\vee \\
{\scriptstyle s\rho} \downarrow & & \downarrow {\scriptstyle j^\vee} \\
s(T, \overline{\partial}) & \xrightarrow{\cong} & (V \cap \ker d_1, d_0)^\vee,
\end{array}
$$

where $j : V \cap \ker d_1 \to V$ *is the natural injection.*

(iv) $H(\overline{\mathbb{L}}_T, \partial) = 0$ *if and only if* $\overline{\mathbb{L}}_T^{(2)}$ *has a closed direct summand of the form* $T = E \oplus \partial E$ *where* $\partial : E \xrightarrow{\cong} \partial E$.

Proof

(i) This is an immediate consequence of the definition of a profree dgl.

(ii) Filtration of $\wedge V$ by (degree $-$ wedge degree) results in a spectral sequence converging to $H(\wedge V, d_0 + d_1)$. Its E_1-term and E_2-term are, respectively, $H(\wedge V, d_1)$ and $(\mathbb{Q} \oplus (V \cap \ker d_1), d_0)$. This proves (ii).

(iii) By Lemma 4.1, the inclusion $j : V \cap \ker d_1 \to V$ is dual to $\overline{\mathbb{L}}_T \to \overline{\mathbb{L}}_T / \overline{\mathbb{L}}_T^{(2)}$, since $(\wedge V, d_1)$ is the quadratic model of $\overline{\mathbb{L}}_T$. This gives (iii).

(iv) If $H(\overline{\mathbb{L}}_T, \partial) = 0$, it follows from the isomorphism $s(\overline{\mathbb{L}}_T, \partial) \cong (V, d_0)^\vee$ that $H(V, d_0) = 0$. This then implies that $(\wedge V, d_0 + d_1)$ is a contractible Sullivan algebra. Therefore $H(V, d_0 + d_1) = \mathbb{Q}$, and by (iii), $H(T, \overline{\partial}) = 0$.

Write $T = E \oplus \overline{\partial} E$, where E and therefore $\overline{\partial} E$ are closed subspaces. Then observe that

$$\overline{\mathbb{L}}_T = \overline{\mathbb{L}}_E \,\widehat{\sqcup}\, \overline{\mathbb{L}}_{\overline{\partial} E}.$$

□

Proposition 12.2

(i) *Let* $\ell : (L, \partial) \to (L', \partial')$ *be a morphism of profree dgl's inducing* $\overline{\ell} : L/L^{(2)} \to L'/(L')^{(2)}$. *Denote the corresponding bihomogeneous morphism of their semi-quadratic Sullivan algebras by*

$$\varphi : (\wedge V, d_0 + d_1) \longleftarrow (\wedge V', d_0' + d_1').$$

Then the following three conditions are equivalent:
(a) $H(\overline{\ell})$ *is an isomorphism.*
(b) $H(\ell)$ *is an isomorphism.*
(c) $H(\varphi)$ *is an isomorphism.*
(ii) *If* L *and* L' *are minimal, then the conditions above are equivalent to each of the conditions*

$$(d) \; \ell \text{ is an isomorphism} \quad (e) \; \varphi \text{ is an isomorphism.}$$

Proof

(i) It follows from the isomorphism $s(L/L^{(2)}) \xrightarrow{\cong} (V \cap \ker d_1, d_0)^\vee$ and Proposition 12.1 (ii) that $H(\overline{\ell})$ is an isomorphism if and only if $H(\varphi)$ is an isomorphism. Moreover, by Proposition 11.3, $H(\ell)$ is an isomorphism if and only if $H(\varphi)$ is an isomorphism. Finally, $\varphi : (\wedge V', d_0 + d_1) \to (\wedge V, d_0 + d_1)$ induces a morphism of minimal Sullivan algebras of the form

$$\psi : (\wedge H(V', d_0), \widehat{d}) \longrightarrow (\wedge H(V, d_0), \widehat{d}).$$

Here the linear map induced from ψ by division by $\wedge^{\geq 2}$ is $H(\varphi, d_0)$. Therefore,

$$H(\varphi, d_0) \text{ is an isomorphism} \iff \psi \text{ is an isomorphism}$$
$$\iff H(\psi) \text{ is an isomorphism}$$
$$\iff H(\varphi) \text{ is an isomorphism.}$$

(ii) If L and L' are minimal, then $\bar{\ell} = H(\bar{\ell})$ is an isomorphism if and only if φ is also an isomorphism. $\qquad\qquad\square$

Proposition 12.3

(i) *An enriched dgl (L, ∂) admits a surjective quasi-isomorphism from a profree dgl $(\mathbb{L}_T, \partial)$.*

(ii) *If $f : (E, \partial) \to (F, \partial)$ is a surjective quasi-isomorphism of enriched dgl's, then any morphism $g : (\mathbb{L}_T, \partial) \to (F, \partial)$ from a profree dgl lifts to provide a commutative diagram*

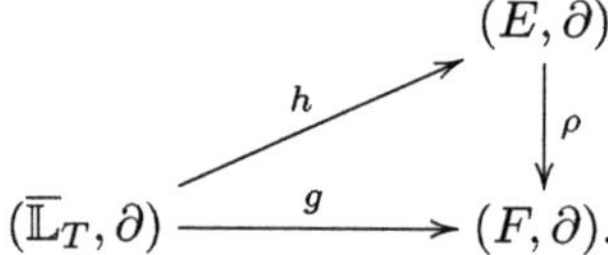

(iii) *Any profree dgl (L, ∂) is the free product $(L, \partial) = (L', \partial') \,\widehat{\amalg}\, (L'', \partial'')$ of a profree dgl satisfying $H(L') = 0$ and a minimal profree dgl (L'', ∂'').*

(iv) *An enriched dgl (L, ∂) admits a unique, up to isomorphism, minimal profree dgl model.*

Proof

(i) Let $(\wedge V, d_0 + d_1)$ be the semi-quadratic model of (L, ∂). We construct by induction on n semi-quadratic Sullivan algebras of the form $(\wedge V \otimes \wedge W(n), d_0 + d_1)$ with $W(0) \subset W(1) \subset \dot{W}(n) \subset W(n+1)\ldots$ and such that:

(1) The induced map $H^{[2]}(\wedge V \otimes \wedge W(n), d_1) \to H^{[2]}(\wedge V \otimes \wedge W(n+1), d_1)$ is zero.

(2) The induced map $H(V \oplus W(n), d_0) \to H(V \oplus W(n+1), d_0)$ is an isomorphism.

Here $H^{[2]}$ denotes as before the component of the cohomology of wedge degree 2.

Suppose now that $\wedge V \otimes \wedge W(n)$ has been defined. Then let $Z = s^{-1}H^{[2]}(\wedge V \otimes \wedge W(n), d_1)$, and set $W(n+1) = W(n) \oplus Z \oplus d_0(Z)$. Then for $x = s^{-1}\Omega \in Z$, choose a d_1-cycle ω with $[\omega] = \Omega$, and define $d_1(x) = \omega$ and $d_1(d_0 x) = -d_0\omega$.

Denote $W = \cup_n W(n)$. Since $H^{[2]}(\wedge V \otimes \wedge W, d_1) = 0$, by Theorem 6.1, $(\wedge V \otimes \wedge W, d_0 + d_1)$ is the semi-quadratic model of a profree dgl $(\mathbb{L}_T, \partial)$, and the injection $\wedge V \to \wedge V \otimes \wedge W$ corresponds to a surjective quasi-isomorphism $(\mathbb{L}_T, \partial) \to (L, \partial)$.

(ii) Since T is closed and ρ is a surjective morphism of enriched vector spaces, there is a map of enriched vector spaces $\alpha : T \to E$ such that $\rho \circ \alpha = g$. Moreover, since ρ is a surjective quasi-isomorphism, $H(\ker \rho) = 0$. Thus there is a morphism of enriched vector spaces $\tau : \ker \rho \cap \ker \partial \to \ker \rho$ such that $\partial \circ \tau$ is the identity in $\ker \rho \cap \ker \partial$.

Now assume h has been defined in $T_{<k}$ and therefore also in $\overline{\mathbb{L}}_{T_{\leq k}}$. Then $h \circ \partial - \partial \circ \alpha : T_k \to \ker \rho \cap \ker \partial$. Extend h to T_k by setting

$$hx = \alpha x + \tau(h\partial - \partial\alpha)(x).$$

It is immediate that the extension to $\overline{\mathbb{L}}_{T_{\leq n}}$ is a morphism and that $\rho \circ h = g$.

(iii) Denote by $\overline{\partial}$ the quotient differential in $L/L^{(2)}$. Then

$$L/L^{(2)} = S \oplus \overline{\partial}S \oplus H,$$

where $S, \overline{\partial}S$, and H are closed subspaces, $\overline{\partial} : S \xrightarrow{\cong} \overline{\partial}S$, and $\overline{\partial}(H) = 0$. Now the isomorphism $T \cong L/L^{(2)}$ from a closed direct summand of $L^{(2)}$ yields a decomposition

$$T = P \oplus \partial P \oplus Q,$$

in which $\partial : P \xrightarrow{\cong} \partial P$ and $\partial : Q \to L^{(2)}$. It follows that $L = \overline{\mathbb{L}}_{(T \oplus \partial P \oplus Q)}$.

Moreover, filtering by Lie bracket length provides a spectral sequence converging from $\overline{\mathbb{L}}_Q$ to $H(L)$. This implies that division by P and ∂P yields a quasi-isomorphism

$$\rho : (L, \partial) \xrightarrow{\cong} (\overline{\mathbb{L}}_Q, \partial_Q),$$

where ∂_Q is the quotient differential.

Finally, by (ii) ρ admits a splitting $\sigma : (\overline{\mathbb{L}}_Q, \partial_Q) \to (L, \partial)$, and it is immediate from the construction that

$$id \,\widehat{\amalg}\, \sigma : \overline{\mathbb{L}}_{P \oplus \partial P)} \,\widehat{\amalg}\, (\overline{\mathbb{L}}_Q, \partial_Q) \xrightarrow{\cong} (L, \partial).$$

Since $\partial : Q \to \mathbb{L}_Q^{(2)}$, this completes the proof.

(iv) Let $\varphi : \overline{\mathbb{L}}_T \to L$ and $\varphi' : \overline{\mathbb{L}}_{T'} \to L$ be minimal dgl models of L. By taking the free product with an acyclic profree dgl $\overline{\mathbb{L}}_Z$, we can suppose that $\varphi' : \overline{\mathbb{L}}_{T'} \,\widehat{\amalg}\, \overline{\mathbb{L}}_Z \to L$ is surjective. Then by (ii) we get a quasi-isomorphism $\theta : \overline{\mathbb{L}}_T \to \overline{\mathbb{L}}_{T'} \,\widehat{\amalg}\, \overline{\mathbb{L}}_Z$ making commutative the diagram

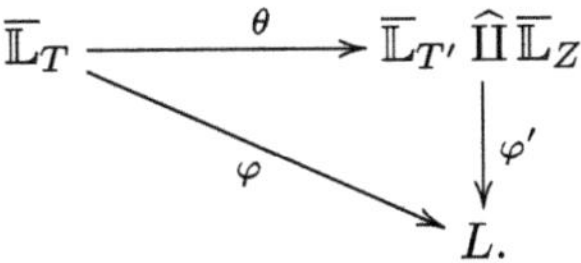

By sending Z to 0 we get a quasi-isomorphism $\theta' : \overline{\mathbb{L}}_T \to \overline{\mathbb{L}}_{T'}$. It follows then from Proposition 11.1 (ii) that the induced map $T \to T'$ is an isomorphism. Therefore θ' is also an isomorphism. $\square$

The Model Category of Enriched dgl's 13

In this section we show that the enriched dgl's inherit a structure of model category. Then we will describe cylinder objects and homotopy of morphisms.

13.1 The Model Category Enrich

Definition In the category of enriched dgl's:

(1) A *cofibration* (or a *profree extension*) is an inclusion of the form $i : (L, \partial) \to (L \,\widehat{\amalg}\, \overline{\mathbb{L}}_S, \partial)$ of enriched dgl's.

  A *trivial profree extension* is a cofibration of the form $L \to L \,\widehat{\amalg}\, \overline{\mathbb{L}}(S \oplus \partial S)$, where $S \xrightarrow{\cong} \partial S$ is an isomorphism.

(2) The *quotient* of a cofibration $(L, \partial) \to (L \,\widehat{\amalg}\, \overline{\mathbb{L}}_S, \partial)$ is the profree dgl $(\overline{\mathbb{L}}_S, \overline{\partial})$ obtained by dividing by the ideal generated by L. The cofibration is *minimal* if the quotient is a minimal profree dgl.

(3) The *spectral sequence* of the cofibration is the spectral sequence obtained by giving to L the gradation 0 and to each element of $\overline{\mathbb{L}}_S$ its degree.

Cofibrations $(L, \partial) \to (L \,\widehat{\amalg}\, \overline{\mathbb{L}}_S, \partial)$ have the following properties:

(i) The spectral sequence converges from $H(L) \,\widehat{\amalg}\, \overline{\mathbb{L}}_S$ to $H(L \,\widehat{\amalg}\, \overline{\mathbb{L}}_S)$.

(ii) The natural map $L \,\widehat{\amalg}\, \overline{\mathbb{L}}_S \to \varprojlim_n L/L^{(n)} \,\widehat{\amalg}\, \overline{\mathbb{L}}_S$ is an isomorphism.

Theorem 13.1 *The category of enriched dgl's is a closed model category where:*

- *The weak equivalences are the quasi-isomorphisms.*

  123
Y. Félix, S. Halperin, *Lie Models for Spaces*, Frontiers in Mathematics,
https://doi.org/10.1007/978-3-032-15357-9_13

- *The fibrations are the surjective morphisms.*
- *The cofibrations are the profree extensions.*

Proof **MC1** Finite limits and colimits for enriched dgl's are constructed in the same way as finite limits and colimits in Sect. 6.7.

MC2 is trivially satisfied and also **MC3** for weak equivalences and fibrations. **MC3** for cofibrations is proved in Proposition 13.1.

MC4 is proved in Proposition 13.2.

MC5 is proved in Proposition 13.3. □

Proposition 13.1 *MC3 for dgl's: The retract of a cofibration is a cofibration.*

Proof Let f be a retract of a cofibration g. This means that we have a commutative diagram

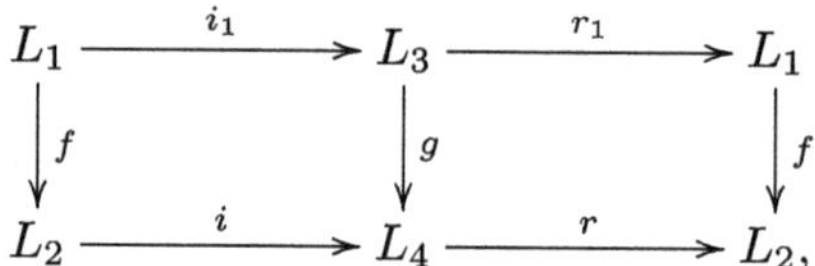

with $r \circ i = id$ and $r_1 \circ i_1 = id$. Note that the subscripts $1, 2, 3, 4$ are not the degrees.

We first consider the situation where $i_1 = r_1 = id_{L_1}$. We write $g : L_1 \to L_1 \,\widehat{\amalg}\, \overline{\mathbb{L}}_T$. We then have the following commutative diagram:

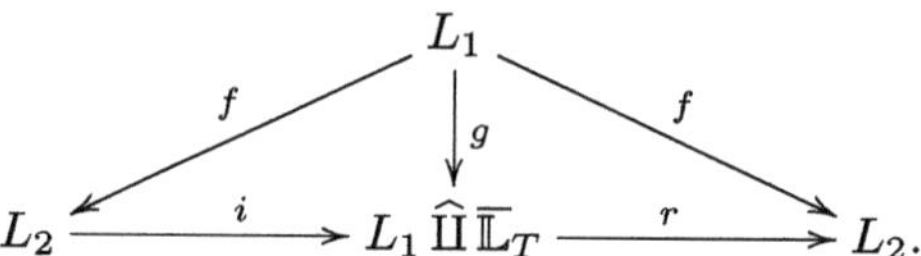

Passing to the indecomposable elements gives

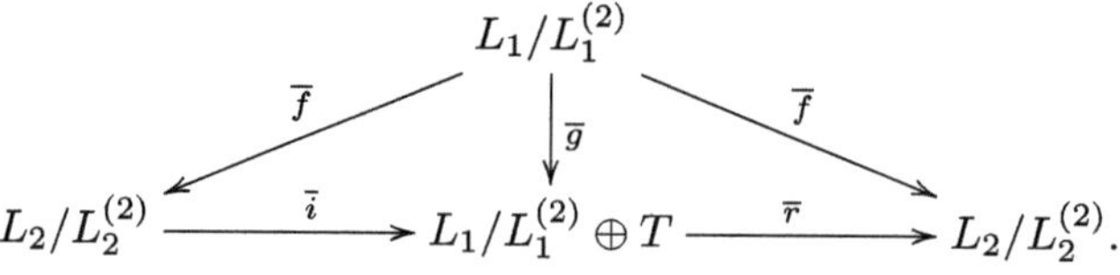

It follows that $\overline{f}$ and f are injective. Denote by S a direct summand of Im $\overline{f}$ in $L_2/L_2^{(2)}$, and choose S so that $\overline{i} : S \to T$. Then $\overline{i} : S \to T$ is injective, and f extends to an isomorphism $h : L_1 \,\widehat{\amalg}\, \overline{\mathbb{L}}_S \to L_2$, from a profree extension.

In the general case, suppose $g : L_3 \to L_4 = L_3 \,\widehat{\amalg}\, \overline{\mathbb{L}}_T$, and take the pushout of g and r_1

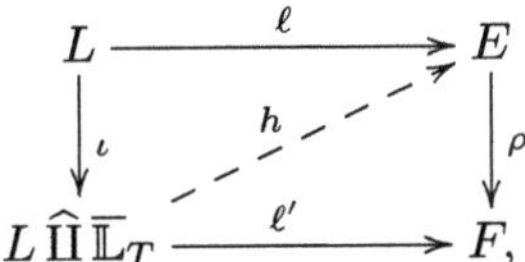

It follows that $L_1 \to L_2$ is a retract of $L_1 \to L_1 \widehat{\amalg} \overline{\mathbb{L}}_T$, and we are reduced to the particular case above. $\square$

Proposition 13.2 *MC4 for enriched dgl's: lifting properties*

Proof The proof contains two parts.

Suppose first to have a commutative diagram

$$
\begin{array}{ccc}
L & \xrightarrow{\ \ell\ } & E \\
\downarrow{\iota} & \overset{h}{\nearrow} & \downarrow{\rho} \\
L\,\widehat{\amalg}\,\overline{\mathbb{L}}_T & \xrightarrow{\ \ell'\ } & F,
\end{array}
$$

in which ι is a cofibration and that ρ is an acyclic fibration. Since T is closed and ρ is a surjective morphism of enriched vector spaces, there is a morphism $\alpha : T \to E$ of enriched vector spaces such that $\rho \circ \alpha = id$: Assume h has been defined in $T_{<k}$ and therefore also in $L\,\widehat{\amalg}\,\overline{\mathbb{L}}_{T_{<k}}$. Let $x \in T_k$. Since $\partial g(x) = g\partial x = \rho(h\partial x)$, $h\partial x = \partial z$ for some z, and we define $h(x) = z$. It is immediate that the extension to $L\,\widehat{\amalg}\,\overline{\mathbb{L}}_{T_{\leq k}}$ is a morphism and that $\rho \circ h = g$.

When $i : L \to L\,\widehat{\amalg}\,\overline{\mathbb{L}}_T$ is an acyclic cofibration, then $H(\overline{\mathbb{L}}_T) = 0$ and $(\overline{\mathbb{L}}_T, \partial) \cong (\overline{\mathbb{L}}_{E \oplus E'}, \partial)$, where $\partial : E \overset{\cong}{\to} E'$ is an isomorphism. In this case we use the surjectivity of ρ to define $h(T)$ such that $\rho \circ h = \ell'$ and then define $h(\partial t) = \partial h(t)$. $\square$

Proposition 13.3 *MC5 for dgl's: decompositions of maps*

Proof Let $\varphi : (E, \partial_E) \to (F, \partial_F)$ be a morphism of enriched dgl's.

Step 1. We first show that φ factors as

$$
\varphi : (E, \partial_E) \xrightarrow{\ \iota\ } (E\,\widehat{\amalg}\,\overline{\mathbb{L}}_T, \partial) \xrightarrow{\ \rho\ } (F, \partial_F)
$$

in which ι is a cofibration and ρ is an acyclic fibration.

From Proposition 12.3 we deduce a surjective quasi-isomorphism $\sigma : (\overline{\mathbb{L}}_{T(0)}, \partial) \overset{\simeq}{\to} (F, \partial_F)$. By Proposition 11.5 (i) this extends to the morphism

$$
\rho(0) = \varphi\,\widehat{\amalg}\,\sigma : (E, \partial_E)\,\widehat{\amalg}\,(\overline{\mathbb{L}}_{T(0)}, \partial) \xrightarrow{\qquad} (F, \partial_F).
$$

By construction, both $\varphi\,\widehat{\amalg}\,\sigma$ and $H(\varphi\,\widehat{\amalg}\,\sigma)$ are surjective.

Next we construct a sequence of cofibrations

$$E \mathbin{\widehat{\amalg}} \overline{\mathbb{L}}_{T(0)} \mathbin{\widehat{\amalg}} L(n-1) \longrightarrow E \mathbin{\widehat{\amalg}} \overline{\mathbb{L}}_{T(0)} \mathbin{\widehat{\amalg}} L(n), \qquad \text{with } L(n) = L(n-1) \mathbin{\widehat{\amalg}} \overline{\mathbb{L}}(Z(n)).$$

These will be equipped with morphisms

$$\rho(n) : E \mathbin{\widehat{\amalg}} \overline{\mathbb{L}}_{T(0)} \mathbin{\widehat{\amalg}} L(n) \longrightarrow F$$

with $\rho(n)$ extending $\rho(n-1)$ and $H_{<n}(\rho(n))$ an isomorphism. Then $\overline{\mathbb{L}}_T = \varinjlim_n (\overline{\mathbb{L}}_{T(0)} \mathbin{\widehat{\amalg}} L(n))$ and the $\rho(n)$ define a quasi-isomorphism

$$\rho : E \mathbin{\widehat{\amalg}} \overline{\mathbb{L}}_T \xrightarrow{\;\simeq\;} F.$$

The construction is by induction on n. If $\rho(n-1)$ is constructed, we may, since $\rho(n-1)$ is a morphism, write

$$\ker \rho(n-1) \cap \ker \partial = C \oplus \partial(\ker \rho(n-1))$$

in which $C = C_{\geq n-1}$ is a closed subspace. Note that since $H(\rho(n-1))$ is surjective, if a cycle x in $\mathrm{Ker}\,\rho(n-1)$ is a boundary, then $a = \partial b$ with $b \in \mathrm{Ker}\,\rho(n-1)$. Thus $C \cong H_{n-1}(\ker \rho(n-1))$. Set $Z(n) = sC_{n-1}$, and define $\partial|_{Z(n)}$ to be the isomorphism $Z(n) \xrightarrow{\;\cong\;} C_{n-1}$. Then extend $\rho(n-1)$ to $\rho(n)$ by setting $\rho(n)(Z(n)) = 0$.

Step 2. Now we show that $f : (L, \partial) \to (L', \partial)$ factors as a composite

$$(L, \partial) \xrightarrow{\;\;i\;\;} (L, \partial) \mathbin{\widehat{\amalg}} (\mathbb{L}_T, \partial) \xrightarrow{\;\;\widetilde{f}\;\;} (L', \partial)$$

where i is an acyclic cofibration and $\widetilde{f}$ a fibration. Let P be an enriched vector space isomorphic with $L'_{\geq 1}$, and let $T = P \oplus \partial P$ with $\partial : P \xrightarrow{\;\cong\;} \partial P$. Use the inclusion $P \to L'$ to extend f to a morphism

$$\widehat{f} : L \mathbin{\widehat{\amalg}} \overline{\mathbb{L}}_T \to L'.$$

Since $H(\overline{\mathbb{L}}_T) = 0$, it follows that the inclusion $i : L \to L \mathbin{\widehat{\amalg}} \overline{\mathbb{L}}_T$ is a quasi-isomorphism. Finally, by construction $\widehat{f}$ is surjective in degrees ≥ 1. $\qquad\qquad\square$

The following propositions establish other basic properties of cofibrations.

Proposition 13.4 *A morphism* $\varphi : (E, \partial_E) \to (F, \partial_F)$ *of enriched dgl's factors as*

$$(E, \partial_E) \xrightarrow{\ \sigma\ } (E \mathbin{\widehat{\amalg}} \overline{\mathbb{L}}_T, \partial) \xrightarrow[\simeq]{\ \pi\ } (F, \partial_F)$$

in which σ is a minimal cofibration and π is a quasi-isomorphism.

In particular, for any enriched dgl F, there is a quasi-isomorphism $\overline{\mathbb{L}}_T \xrightarrow{\ \simeq\ } F$ from a profree dgl.

Proof By MC2 we get a decomposition of φ

$$E \to E \mathbin{\widehat{\amalg}} \overline{\mathbb{L}}_T \xrightarrow{\ \rho\ } F$$

in which ρ is a quasi-isomorphism.

Denote by $\overline{\partial}$ the quotient differential in T. Then

$$(T, \overline{\partial}) = S \oplus \overline{\partial} S \oplus Q,$$

where $S, \overline{\partial} S$, and Q are closed subspaces, $\overline{\partial} : S \xrightarrow{\cong} \overline{\partial} S$, and $\overline{\partial}(Q) = 0$. Filtering by the Lie bracket length provides then a spectral sequence which shows that this map is a quasi-isomorphism. This also implies that division by S and ∂S yields a quasi-isomorphism

$$q : (L \mathbin{\widehat{\amalg}} \overline{\mathbb{L}}_T, \partial) \xrightarrow{\cong} (L \mathbin{\widehat{\amalg}} \overline{\mathbb{L}}_Q, \partial_Q)$$

where ∂_Q is the quotient differential.

Finally, by CM4 ρ admits a section $\sigma : (L \mathbin{\widehat{\amalg}} \overline{\mathbb{L}}_Q, \partial_Q) \to (L \mathbin{\widehat{\amalg}} \overline{\mathbb{L}}_T, \partial)$, and it is immediate from the construction that

$$q = \sigma \mathbin{\widehat{\amalg}} \psi : (L \mathbin{\widehat{\amalg}} \overline{\mathbb{L}}_Q), \partial_Q) \mathbin{\widehat{\amalg}} (\overline{\mathbb{L}}_{S \oplus \partial S}) \xrightarrow{\cong} (L \mathbin{\widehat{\amalg}} \overline{\mathbb{L}}_T, \partial),$$

where γ is the restriction of ψ.

Then restrict q to $E \mathbin{\widehat{\amalg}} \overline{\mathbb{L}}_Q$ to get the decomposition

$$E \xrightarrow{\ q \circ j\ } E \mathbin{\widehat{\amalg}} \overline{\mathbb{L}}_Q \xrightarrow{\ q \circ \sigma\ } F.$$

$\square$

Proposition 13.5 *Suppose that*

$$\begin{array}{ccccc}
E & \longrightarrow & E \mathbin{\widehat{\amalg}} \overline{\mathbb{L}}_S & \xrightarrow[\simeq]{\ \alpha'\ } & F \\
{\scriptstyle \gamma_E}\downarrow & & & & \downarrow{\scriptstyle \gamma'} \\
E' & \longrightarrow & E' \mathbin{\widehat{\amalg}} \overline{\mathbb{L}}_{S'} & \xrightarrow[\simeq]{\ \alpha\ } & F'
\end{array}$$

is a commutative diagram of cofibrations, and suppose that α is an acyclic fibration. Then there is a morphism

$$\gamma : E \,\widehat{\amalg}\, \overline{\mathbb{L}}_S \to E' \,\widehat{\amalg}\, \overline{\mathbb{L}}_{S'}$$

for which the diagram

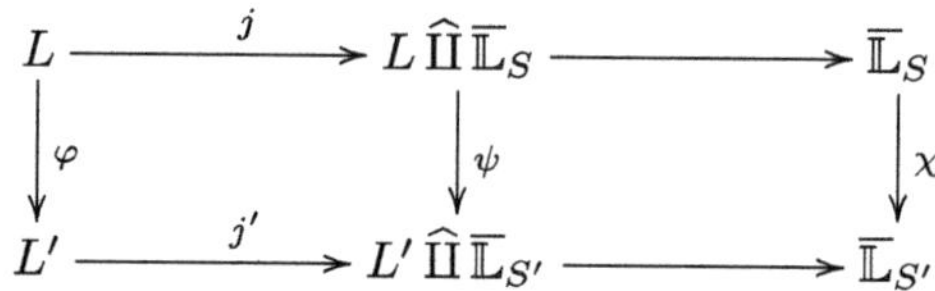

commutes.

Proof Assume by induction that γ is defined in $S_{<n}$, and let x be a basis element in S_n. Then $\partial x \in E \,\widehat{\amalg}\, \overline{\mathbb{L}}_{S_{<n}}$ and

$$\alpha\gamma\,\partial x = \partial\gamma'\alpha' x.$$

Since $\gamma\,\partial x$ is a cycle and α is a surjective quasi-isomorphism, there is $w \in E' \,\widehat{\amalg}\, \overline{\mathbb{L}}_{S'}$ such that $\alpha w = \gamma'\alpha'(x)$ and $\gamma\,\partial x = \partial w$. We extend γ by setting $\gamma x = w$. $\square$

Proposition 13.6 *Suppose*

$$
\begin{array}{ccccc}
L & \xrightarrow{\;j\;} & L\,\widehat{\amalg}\,\overline{\mathbb{L}}_S & \longrightarrow & \overline{\mathbb{L}}_S \\
\downarrow{\scriptstyle\varphi} & & \downarrow{\scriptstyle\psi} & & \downarrow{\scriptstyle\chi} \\
L' & \xrightarrow{\;j'\;} & L'\,\widehat{\amalg}\,\overline{\mathbb{L}}_{S'} & \longrightarrow & \overline{\mathbb{L}}_{S'}
\end{array}
$$

is a commutative diagram connecting two cofibrations. If any two of φ, ψ, and χ are quasi-isomorphisms, then so is the third.

Proof We proceed in a number of steps.
Step 1. *If $\overline{\mathbb{L}}_S = \overline{\mathbb{L}}_{S'}$ and $\psi = \varphi\,\widehat{\amalg}\,id_{\overline{\mathbb{L}}_S}$, then ψ is a quasi-isomorphism if and only if φ is a quasi-isomorphism.*

If φ is a quasi-isomorphism, it follows from the associated spectral sequence that $\varphi\,\widehat{\amalg}\,id_{\overline{\mathbb{L}}_S}$ is a quasi-isomorphism.

In the reverse direction, suppose $\varphi\,\widehat{\amalg}\,id_{\overline{\mathbb{L}}_S}$ is a quasi-isomorphism. Factor φ as (Proposition 13.4)

$$L \to L\,\widehat{\amalg}\,\overline{\mathbb{L}}_T \xrightarrow{\;\simeq\;} L'$$

in which $L\,\widehat{\amalg}\,\overline{\mathbb{L}}_T$ is a minimal profree extension. This then extends to

$$L\,\widehat{\amalg}\,\overline{\mathbb{L}}_S \to L\,\widehat{\amalg}\,\overline{\mathbb{L}}_T\,\widehat{\amalg}\,\overline{\mathbb{L}}_S \xrightarrow{\simeq} L'\,\widehat{\amalg}\,\overline{\mathbb{L}}_S,$$

in which the second morphism is a quasi-isomorphism. Since the hypothesis $L\,\widehat{\amalg}\,\overline{\mathbb{L}}_S \to L'\,\widehat{\amalg}\,\overline{\mathbb{L}}_S$ is also a quasi-isomorphism, it follows that

$$\psi : L\,\widehat{\amalg}\,\overline{\mathbb{L}}_S \to L\,\widehat{\amalg}\,\overline{\mathbb{L}}_S\,\widehat{\amalg}\,\overline{\mathbb{L}}_T = L\,\widehat{\amalg}\,\overline{\mathbb{L}}_T\,\widehat{\amalg}\,\overline{\mathbb{L}}_S$$

is also a quasi-isomorphism.

Suppose $T = T_{\geq n}$, and let x be an element of T_n. Then, since ψ is a quasi-isomorphism, $\partial x = \partial a$ for some $a \in L\,\widehat{\amalg}\,\overline{\mathbb{L}}_S$. Then $x - a$ is a cycle, and there are $b \in L\,\widehat{\amalg}\,\overline{\mathbb{L}}_S$ and y in $L\,\widehat{\amalg}\,\overline{\mathbb{L}}_S\,\widehat{\amalg}\,\overline{\mathbb{L}}_T$ such that $x - a = b + \partial y$. This contradicts the minimality of the extension $L \to L\,\widehat{\amalg}\,\overline{\mathbb{L}}_T$.

Step 2. *If φ is a quasi-isomorphism, then ψ is a quasi-isomorphism if and only if χ is a quasi-isomorphism.*

Proposition 13.4 provides a quasi-isomorphism $L\,\widehat{\amalg}\,\overline{\mathbb{L}}_T \xrightarrow{\simeq} 0$ from a minimal profree extension. If φ is a quasi-isomorphism, it follows from Step 1 that the extension $\varphi' : L\,\widehat{\amalg}\,\overline{\mathbb{L}}_T \xrightarrow{\simeq} L'\,\widehat{\amalg}\,\overline{\mathbb{L}}_T$ is a quasi-isomorphism and so $H(L'\,\widehat{\amalg}\,\overline{\mathbb{L}}_T) = 0$. This implies that the inclusions

$$L\,\widehat{\amalg}\,\overline{\mathbb{L}}_T \longrightarrow L\,\widehat{\amalg}\,\overline{\mathbb{L}}_T\,\widehat{\amalg}\,\overline{\mathbb{L}}_S \quad \text{and} \quad L'\,\widehat{\amalg}\,\overline{\mathbb{L}}_T \longrightarrow L'\,\widehat{\amalg}\,\overline{\mathbb{L}}_T\,\widehat{\amalg}\,\overline{\mathbb{L}}_{S'}$$

extend to isomorphisms

$$(L\,\widehat{\amalg}\,\overline{\mathbb{L}}_T, \partial)\,\widehat{\amalg}\,(\overline{\mathbb{L}}_S, \partial_S) \xrightarrow{\cong} L\,\widehat{\amalg}\,\overline{\mathbb{L}}_T\,\widehat{\amalg}\,\overline{\mathbb{L}}_S = L\,\widehat{\amalg}\,\overline{\mathbb{L}}_S\,\widehat{\amalg}\,\overline{\mathbb{L}}_T$$

and

$$(L'\,\widehat{\amalg}\,\overline{\mathbb{L}}_T, \partial)\,\widehat{\amalg}\,(\overline{\mathbb{L}}_{S'}, \partial_{S'}) \xrightarrow{\cong} L'\,\widehat{\amalg}\,\overline{\mathbb{L}}_T\,\widehat{\amalg}\,\overline{\mathbb{L}}_{S'} = L'\,\widehat{\amalg}\,\overline{\mathbb{L}}_{S'}\,\widehat{\amalg}\,\overline{\mathbb{L}}_T$$

Now by Step 1, ψ is a quasi-isomorphism if and only if $\psi\,\widehat{\amalg}\,id_{\overline{\mathbb{L}}_T}$ is a quasi-isomorphism. The result follows then from the commutative diagram

$$
\begin{array}{ccc}
(L\,\widehat{\amalg}\,\overline{\mathbb{L}}_T)\,\widehat{\amalg}\,\overline{\mathbb{L}}_S & \xrightarrow{\;\simeq\;} & L\,\widehat{\amalg}\,\overline{\mathbb{L}}_T\,\widehat{\amalg}\,\overline{\mathbb{L}}_S \\
\downarrow{\scriptstyle \psi'\,\widehat{\amalg}\,\overline{\psi}} & & \downarrow{\scriptstyle \psi\,\widehat{\amalg}\,id_{\overline{\mathbb{L}}_T}} \\
(L'\,\widehat{\amalg}\,\overline{\mathbb{L}}_T)\,\widehat{\amalg}\,\overline{\mathbb{L}}_{S'} & \xrightarrow{\;\simeq\;} & L'\,\widehat{\amalg}\,\overline{\mathbb{L}}_T\,\widehat{\amalg}\,\overline{\mathbb{L}}_{S'}.
\end{array}
$$

Step 3. *Completion of the proof.*

Here we may assume that ψ and χ are quasi-isomorphisms, and we have to show that φ is a quasi-isomorphism. Now Proposition 13.4 extends φ to a quasi-isomorphism

$$\alpha : L \mathbin{\widehat{\amalg}} \overline{\mathbb{L}}_U \xrightarrow{\;\simeq\;} L'$$

from a minimal profree extension. We then deduce a map

$$\beta : L \mathbin{\widehat{\amalg}} \overline{\mathbb{L}}_U \mathbin{\widehat{\amalg}} \overline{\mathbb{L}}_S \to L' \mathbin{\widehat{\amalg}} \overline{\mathbb{L}}_{S'}$$

with $\beta = j' \circ \alpha$ on $L \mathbin{\widehat{\amalg}} \overline{\mathbb{L}}_U$ and $\beta = \psi$ on $L \mathbin{\widehat{\amalg}} \overline{\mathbb{L}}_S$. This gives the commutative diagram

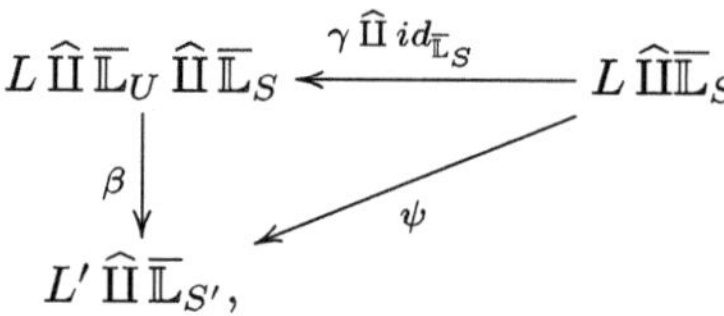

in which γ is the inclusion $L \to L \mathbin{\widehat{\amalg}} \overline{\mathbb{L}}_U$. Since α and χ are quasi-isomorphisms, it follows from Step 2 that β is a quasi-isomorphism. Since ψ is a quasi-isomorphism, so is $\gamma \mathbin{\widehat{\amalg}} id_{\overline{\mathbb{L}}_S}$. By Step 1, γ is also a quasi-isomorphism. It is straightforward to see that, since $L \mathbin{\widehat{\amalg}} \overline{\mathbb{L}}_U$ is a minimal extension, this implies that $U = 0$ and hence that φ is a quasi-isomorphism. $\square$

13.2   Cylinder Objects and dgl Homotopy

A cylinder object is defined here as in any closed model category.

Definition

1. A *cylinder object* for an enriched dgl (L, ∂) is a commutative diagram of enriched dgl morphisms:

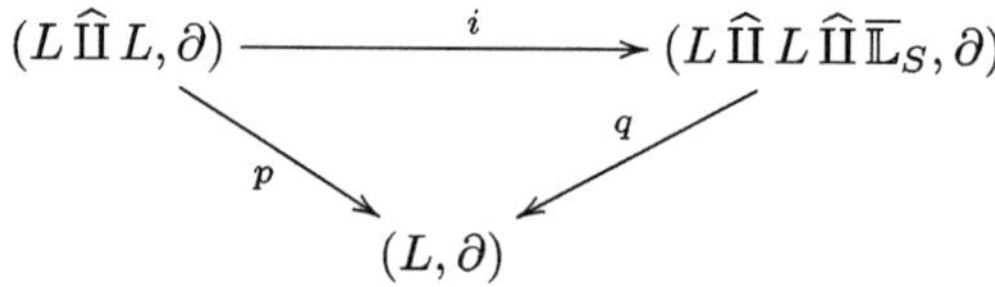

with the following properties:

- i is a cofibration.
- p is the identity in $L \mathbin{\widehat{\amalg}} 0$ and in $0 \mathbin{\widehat{\amalg}} L$.
- q is an acyclic fibration and $q(S) = 0$.

2. Two morphisms f_1, $f_2 : (L, \partial) \to (L', \partial')$ of enriched dgl's are *homotopic* if $f_1 \,\widehat{\amalg}\, f_2 :$ $L \,\widehat{\amalg}\, L \to L'$ extend to a morphism (called the *homotopy*),

$$L \,\widehat{\amalg}\, L \,\widehat{\amalg}\, \overline{\mathbb{L}}_S \to L',$$

from a cylinder object.

Proposition 13.7

 (i) *Any enriched dgl (L, ∂) has a cylinder object.*
 (ii) *Homotopy from a cofibrant object is an equivalence relation and is independent of the choice of cylinder object.*
 (iii) *Homotopy from a cofibrant object is preserved by composition.*
 (iv) *If (L, ∂) is a profree dgl and $h : (L', \partial') \to (L'', \partial'')$ is a quasi-isomorphism of enriched dgl's, then composition with h induces a bijection*

$$[L, L'] \xrightarrow{\;\cong\;} [L, L''].$$

Proof (i) This follows directly from MC2 applied to $p : L \,\widehat{\amalg}\, L \to L$.

   (ii) and (iii) are particular cases of Lemmas 4 and 6 of [73].

   (iv) Decompose $L' \to L''$ into a trivial cofibration $L' \,\widehat{\amalg}\, \overline{\mathbb{L}}_S$ followed by a surjection $L' \,\widehat{\amalg}\, \overline{\mathbb{L}}_S \to L''$. We obtain in this way two acyclic fibrations $L' \,\widehat{\amalg}\, \overline{\mathbb{L}}_S \to L'$ and $L' \,\widehat{\amalg}\, \overline{\mathbb{L}}_S \to L''$. By [73, Lemma 7] we deduce isomorphisms $[L, L'] \cong [L, L' \,\widehat{\amalg}\, \overline{\mathbb{L}}_S] \cong [L, L'']$. $\qquad\square$

Corollary 13.1 *A quasi-isomorphism between minimal profree dgl's is an isomorphism.*

Proof This follows immediately from Proposition 12.2. $\qquad\square$

Corollary 13.2 *Let $\varphi : L \to L'$ be a quasi-isomorphism of profree dgl's. Then there is $\psi : L' \to L$ such that $\psi \circ \varphi \sim id$.*

Proof By Proposition 13.7 (iv), composition with φ induces a bijection of sets of homotopy classes

$$[L', L] \to [L, L].$$

There is then ψ with $\psi \circ \varphi \sim id$. $\qquad\square$

In summary we have the following theorem:

Theorem 13.2

(i) Each enriched dgl, (L, ∂), has a minimal profree dgl model.

(ii) Two profree dgl models $f_1, f_2 : \overline{\mathbb{L}}_{T(1)}, \overline{\mathbb{L}}_{T(2)} \xrightarrow{\simeq} L$ determine a unique homotopy class of homotopy equivalences

$$g : \overline{\mathbb{L}}_{T(1)} \to \overline{\mathbb{L}}_{T(2)}$$

such that $f_2 \circ g \sim f_1$. If the models are minimal, then g is an isomorphism.

(iii) More generally, suppose $f_1 : \overline{\mathbb{L}}_{T(1)} \to L(1)$ and $f_2 : \overline{\mathbb{L}}_{T(2)} \to L(2)$ are profree dgl models; then any morphism

$$h : L(1) \to L(2)$$

of enriched dgl's determines a unique homotopy class of morphisms $\widehat{h} : \overline{\mathbb{L}}_{T(1)} \to \overline{\mathbb{L}}_{T(2)}$ such that

$$f_2 \circ \widehat{h} \sim h \circ f_1.$$

$$
\begin{array}{ccc}
\overline{\mathbb{L}}_{T(1)} & \xrightarrow{\ f_1\ } & L_1 \\
\Big\downarrow{\widehat{h}} & & \Big\downarrow{h} \\
\overline{\mathbb{L}}_{T(2)} & \xrightarrow{\ f_2\ } & L_2.
\end{array}
$$

Proof

(i) This is Proposition 13.4.

(ii) We factorize f_2 as the composition of an acyclic cofibration followed by a fibration (Proposition 13.3)

$$\overline{\mathbb{L}}_{T(2)} \xrightarrow{\ j\ } \overline{\mathbb{L}}_{T(2)\oplus S} \xrightarrow{\ \psi_2\ } L.$$

Then by 12.3 there is a map g making commutative the diagram

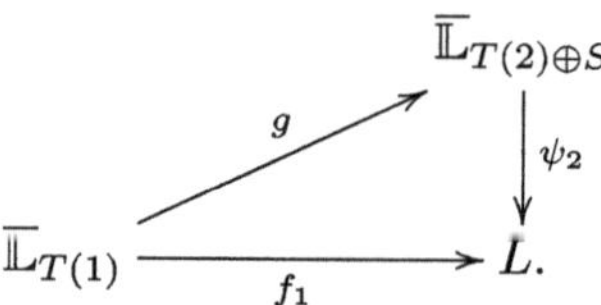

Since g is a quasi-isomorphism of profree dgl's, by Proposition 13.7 (iv)

$$[\overline{\mathbb{L}}_{T(1)}, \overline{\mathbb{L}}_{T(2)}] \to [\overline{\mathbb{L}}_{T(1)}, \overline{\mathbb{L}}_{T(2)\oplus s}]$$

is a bijection. There is thus $g' : \overline{\mathbb{L}}_{T(1)} \to \overline{\mathbb{L}}_{T(2)}$ such that $jg' \sim g$. Then

$$\varphi_2 g' = \psi_2 jg' \sim \psi_2 g' = \varphi_1.$$

(iii) We factorize $h \circ f_1$ as a cofibration followed by an acyclic fibration

$$hf_1 : \overline{\mathbb{L}}_{T(1)} \xrightarrow{\;j\;} \overline{\mathbb{L}}_{T(1)\oplus R} \xrightarrow{\;\psi\;} L_2.$$

By (ii) there is then a map $g : \overline{\mathbb{L}}_{T(1)\oplus R} \to \overline{\mathbb{L}}_{T(2)}$ such that $f_2 g \sim \psi$. We define $\widetilde{h} = g \circ j$. $\qquad\qquad\square$

13.3 Uniqueness of Lie Models and Rationalization

Rational homotopy begins with the work of Quillen [72]. In [72] Quillen constructed a pair of functors between the category of 1-connected rational simplicial sets with finite type homology and the category of 1-connected finite type dgl's

$$\lambda : \qquad \mathrm{sSet}_1 \; \underset{\longleftarrow}{\overset{\longrightarrow}{}} \; \mathrm{dgl}_1 \qquad \langle . \rangle_Q$$

Both the Quillen functor λ and the Quillen realization functor $\langle . \rangle_Q$ are obtained by composition of other functors. In the end this leads to an equivalence of homotopy categories.

The cochain functor associates with any connected finite type dgl a Sullivan algebra. The main result of Majewski was the proof that the cochain algebra on the Quillen model of a space X is in a natural way quasi-isomorphic to the Sullivan minimal model of X [63]. This leads to an equivalence of model categories between the category of connected finite type dgl's and the category of simply connected cdga's with finite type homology.

$$C^* : \mathrm{dgl}_1 \; \underset{\longleftarrow}{\overset{\longrightarrow}{}} \; \mathrm{cdga}_1.$$

More recently the construction by Lawrence and Sullivan of a Lie model for the interval [56] was the starting point of the construction by Buijs et al. [15] of a dgl's model $\mathcal{L}_X$ for any space X. This Lie algebra $\mathcal{L}_X$ is called a *complete dgl*

$$\mathcal{L}_X = \varprojlim_{n} \mathcal{L}_X / \mathcal{L}_X^n.$$

This induces a pair of adjoint functors, global model, and realization

$$\mathcal{L} : \mathrm{sSet} \rightleftarrows \mathrm{cdgl's} \qquad : \langle . \rangle_B$$

which forms a Quillen pair. In particular they induce adjoint functors in the homotopy categories

$$H_0 - \mathrm{sSet} \xrightarrow[\langle \cdot \rangle_B]{\mathcal{L}} H_0 - \mathrm{cdgl's}$$

and both preserve weak equivalences and homotopies.

The complete dgl's models are partially related to the Sullivan models. When $H_*(X; \mathbb{Q})$ is a finite type graded vector space, then $\varinjlim_n C^*(\mathcal{L}_X / \mathcal{L}_X^n)$ is quasi-isomorphic to the Sullivan model of X [15, Theorem 10.8]. This combined with Theorem 11.1 shows that the minimal profree model of a space X with finite type homology is isomorphic to its minimal cdgl model.

In [58] the authors use another realization functor $\langle . \rangle_B$, known as the Deligne-Getzler-Hinich ∞-groupoid. This is the simplicial set of classes of Maurer-Cartan elements in the simplicial Lie algebra $A_{PL}(\Delta^n) \widehat{\otimes} L$ [37, 49].

The relation between the two realizations is described in [15, Theorem 11.14]. Let X be a connected space with finite type homology, and denote by $\overline{L}_X$ its minimal enriched model and by $(\wedge V, d)$ its minimal Sullivan model; then we have a homotopy equivalence

$$\langle \overline{L}_X \rangle_B \cong \langle \wedge V \rangle_{\mathbb{Q}}.$$

When the homology of the space X is not finite type, the complete dgl model $\mathcal{L}_X$ and the profree dgl model $\overline{L}_X$ can be very different and the same for their geometric realizations. Consider for instance the case of an infinite wedge of spheres S^2

$$X = \vee_{i=1}^{\infty} S^2.$$

Its complete Lie model is the free Lie algebra $\mathbb{L}(x_1, \dots)$, and its geometric realization is the wedge of rational spheres, $\vee_{i=1}^{\infty} S_{\mathbb{Q}}^2$.

On the other hand its enriched model L_X can be derived from its Sullivan minimal model $(\wedge V, d)$. Since $H_2(X; \mathbb{Q})$ is an infinite vector space and $V^2 = H^2(X; \mathbb{Q}) \cong (H_2)^{\vee}$, thus,

$$\pi_2(X_{\mathbb{Q}}) \cong (V^2)^{\vee} = ((H_2(X; \mathbb{Q})^{\vee})^{\vee})^{\vee}$$

which is clearly not isomorphic to $H_2(X; \mathbb{Q})$.

The Profree dgl Model of a cdga and of a Topological Space

14

Here we describe the properties of the profree dgl model of a commutative differential graded algebra and of a connected space X. We give explicit examples, in particular for products, wedges, cofibers, and quotients of spaces.

14.1 The Profree dgl Model of a cdga

Definition

(i) A *profree dgl model of a cdga*, (A, d), is a profree dgl $(L = \overline{\mathbb{L}}_T, \partial)$ together with a *connecting quasi-isomorphism*

$$\psi : (\wedge V_L, d_0 + d_1) \xrightarrow{\simeq} (A, d)$$

from the semi-quadratic Sullivan model of $(\overline{\mathbb{L}}_T, \partial)$.

(ii) A *profree dgl representative of a cdga morphism* $\varphi : (A, d) \to (A', d')$ is a morphism $\ell : (L', \partial') \to (L, \partial)$ between respective profree dgl models for which the diagram

$$
\begin{array}{ccc}
(\wedge V_L, d) & \xrightarrow{\ \varphi_\ell\ } & (\wedge V_{L'}, d') \\
\downarrow & & \downarrow \\
(A, d) & \xrightarrow{\ \ \varphi\ \ } & (A', d')
\end{array}
$$

is homotopy commutative.

Now recall that a Sullivan model of a cdga (A, d) is a quasi-isomorphism $(\wedge V, d) \xrightarrow{\simeq} (A, d)$ in which:

- $\wedge V$ is a free graded commutative algebra with $V = V^{\geq 1}$.
- $\wedge V = \varinjlim_n \wedge V(n)$ with $d : V(n+1) \to \wedge V(n)$.

In parallel, a profree dgl model $(\overline{\mathbb{L}}_T, \partial)$ of (A, d) is a quasi-isomorphism $(\wedge V_L, d_1 + d_0) \xrightarrow{\simeq} (A, d)$ in which:

- $L = \overline{\mathbb{L}}_T$ is a profree graded Lie algebra.
- $T = \varinjlim_n T(n)$ with $\partial : T(n+1) \to \overline{\mathbb{L}}_{T(n)}$.

Theorem 14.1

(i) Any cdga (A, d) with $H^0(A) = \mathbb{Q}$ has a profree dgl model (L, ∂) with connecting quasi-isomorphism $\psi : (\wedge V_L, d_0 + d_1) \xrightarrow{\simeq} (A, d)$.

(ii) Let $\psi' : (\wedge V_{L'}, d'_0 + d'_1) \xrightarrow{\simeq} (A', d')$ be a connecting quasi-isomorphism for a profree dgl model, (L', ∂'), of a cdga (A', d'). Then a morphism $f : (A, d) \to (A', d')$ has a profree dgl representative $g : L' \to L$: There is a homotopy commutative diagram

$$
\begin{array}{ccc}
\wedge V_L & \xrightarrow[\simeq]{\psi} & A \\
\downarrow{\scriptstyle \varphi_g} & & \downarrow{\scriptstyle f} \\
\wedge V_{L'} & \xrightarrow[\simeq]{\psi'} & A'.
\end{array}
$$

Definition The morphism

$$g : (L', \partial') \to (L, \partial)$$

corresponding to f is a *profree dgl representative* of f.

Theorem 14.1 follows from the next proposition. We first establish it and then provide the proof of the theorem.

Proposition 14.1

(i) Let (A, d) be a connected cdga, and denote $\overline{A} = A^{\geq 1}$. Then (A, d) admits a natural semi-quadratic model $(\wedge V, d)$ where $(\wedge V, d_0 + d_1)$ is the semi-quadratic algebra associated with a profree dgl $(\overline{\mathbb{L}}_{s^{-1}\overline{A}^\vee}, \partial)$, where:

- $\partial_0 = s^{-1}\overline{A}^\vee \to s^{-1}\overline{A}^\vee$.
- $\partial_1 : s^{-1}\overline{A}^\vee \to \mathbb{L}^2(s^{-1}\overline{A}^\vee)$.
- *For $x \in s^{-1}\overline{A}^\vee$, $\partial(x) = 0$ if sx vanishes on the products $\overline{A} \cdot \overline{A}$.*

(ii) With the same notation we have a commutative diagram

$$
\begin{array}{ccc}
s(\overline{\mathbb{L}}_T, \partial) & \xrightarrow{\;\cong\;} & (V, d_0)^\vee \\
\downarrow{\scriptstyle s\rho} & & \downarrow{\scriptstyle j^\vee} \\
s(T, \partial) & \xrightarrow{\;\cong\;} & (V \cap \ker d_1, d_0)^\vee,
\end{array}
$$

where ρ denotes the projection and $j : V_0 = V \cap \ker d_1 \to V$ is the natural injection.

Proof

(i) The construction is in three steps.

Step 1. *We suppose that $d_A = 0$ and $A^{\geq 1} \cdot A^{\geq 1} = 0$.*

In this case (Proposition 8.6) $(A, 0)$ has a quadratic Sullivan model of the form

$$\psi : (\wedge V, d_1) \xrightarrow{\;\simeq\;} (A, 0)$$

in which $\psi : V \cap \ker d_1 \xrightarrow{\;\cong\;} A^{\geq 1}$. In particular, the homotopy Lie algebra of $(\wedge V, d_1)$ is the profree Lie algebra $\overline{\mathbb{L}}_{s^{-1}(A^{\geq 1})^\vee}$.

Additionally V admits a second gradation, $V = \oplus_{n \geq 0} V_n$, satisfying

$$V_0 = V \cap \ker d_1, \qquad d_1 : V_{n+1} \to (\wedge^2 V)_n, n \geq 0, \qquad \text{and } \psi(V_{\geq 1}) = 0.$$

The construction is by induction on n. First we choose a graded subspace $V_1 \subset V$ for which

$$d_1 : V_1 \xrightarrow{\;\cong\;} \wedge^2(V_0).$$

Then if V_k, $k \leq n$ is defined, we choose a graded subspace $V_{n+1} \subset V$ so that d_1 induces an isomorphism

$$d_1 : V_{n+1} \xrightarrow{\;\cong\;} H_n^2(\wedge V_{\leq n}, d_1).$$

Step 2. *The model of the cdga, $(A, 0)$*

We construct a derivation d_0 in $\wedge V$ such that $d_0 : V_n \to V_{n-1}$, $(d_0 + d_1)^2 = 0$ and such that ψ remains a quasi-isomorphism

$$\psi : (\wedge V, d_0 + d_1) \xrightarrow{\simeq} (A, 0).$$

The construction is by induction on n. First, let $x \in V_1$. Since $\psi : V_0 \xrightarrow{\cong} \overline{A}$ is a linear isomorphism, there is an element $u \in V_0$ such that $\psi(d_1 x) = \psi(u)$. We define $d_0(x) = -u$, and so $\psi(d_0 + d_1)(x) = 0 = d\psi(x)$.

When $x \in V_2$, then $d_0 d_1(x) \in \wedge^2 V_0$ and so there is an element $u \in V_1$ such that $d_0 d_1(x) = -d_1(u)$. We set $d_0(x) = u$, and so $d_1 d_0 + d_0 d_1 = 0$ on V_2. On the other hand, by construction

$$\psi(d_0^2(x)) = -\psi(d_1 d_0(x)) = \psi(d_0 d_1(x)) = -\psi(d_1^2(x)) = 0.$$

Since $\psi : V_0 \to \overline{A}$ is an isomorphism, $d_0^2(x) = 0$.

Suppose now we have defined d_0 on $V_{\leq n}$ for some $n \geq 2$ such that $d_0 d_1 + d_1 d_0 = d_0^2 = 0$. Then we construct d_0 on V_{n+1} as follows: If $x \in V_{n+1}$, then $d_1 d_0 d_1(x) = d_0 d_1^2(x) = 0$. Since $H_{\geq 1}(\wedge V, d_1) = 0$, there is an element $u \in V_n$ such that $d_1 d_0(x) = -d_1(u)$. We define then $d_0(x) = u$. Now $d_0^2(x)$ is a d_1-cycle in V_{n-1}, so a d_1-boundary, which implies that $d_0^2(x) = 0$.

Step 3. *The general case*

We construct a derivation δ in $\wedge V$ restricting to a linear map $\delta : V_n \to V_n, n \geq 0$, and we show that

$$\psi : (\wedge V, d_1 + d_0 + \delta) \to (A, d_A)$$

is a cdga quasi-isomorphism.

The construction of δ and the verification of the properties $\delta^2 = \delta d_0 + d_0 \delta = \delta d_1 + d_1 \delta = 0$ are by induction on n.

On V_0 the differential δ is constructed so that $\psi : (V_0, \delta) \to (\overline{A}, d_A)$ is an isomorphism of complexes. Since $d_1 : V_1 \to \wedge^2 V_0$ is an isomorphism, for each $x \in V_1$, there is an element u such that $\delta d_1(x) = -d_1(u)$. We define $\delta(x) = u$, and we have $\delta d_1 + d_1 \delta = 0$.

Now since $\varphi(d_0) = -\psi(d_1)$, we have

$$\psi(d_0 \delta) = -\psi(d_1 \delta) = \psi(\delta d_1) = d_A \psi d_1 = -d_A \psi d_0 = -\psi(\delta d_0).$$

Therefore $\psi(d_0 \delta + \delta d_0) = 0$ and so $d_0 \delta + \delta d_0 = 0$.

Finally $d_1 \delta^2(x) = 0$, and since $\delta^2(x) \in V_1$ is a d_1-cycle, $\delta^2(x) = 0$.

Now let $x \in V_n, n \geq 2$. By induction $d_1 \delta d_1 = -\delta d_1^2 = 0$, and so there is an element $u \in V_n$ such that $d_1(u) = -\delta d_1(x)$. We set $\delta(x) = u$. Then $\delta^2(x)$ and $d_0 \delta + \delta d_0$ are d_1 cycles in $V_{\geq 1}$ and so there is 0.

Finally, by construction $\psi d_0(V_1) \subset \overline{A} \cdot \overline{A}$.

(ii) This follows from Proposition 12.1 (iii).   □

Proof of Theorem 14.1

(i) follows directly from Proposition 14.1.

(ii) Let $(\wedge V_{L'}, d_0' + d_1') \xrightarrow{\simeq} (A', d')$ be a connecting quasi-isomorphism from the semi-quadratic Sullivan model of a profree dgl. Then there is a homotopy commutative diagram

$$
\begin{array}{ccc}
(\wedge V_L, d_0 + d_1) & \xrightarrow{\;\simeq\;} & (A, d) \\
{\scriptstyle \simeq}\downarrow{\scriptstyle \psi} & & \downarrow{\scriptstyle \varphi_A} \\
(\wedge V_{L'}, d_0' + d_1') & \xrightarrow{\;\simeq\;} & (A', d').
\end{array}
$$

But now Proposition 11.4 asserts that ψ is homotopic to a bihomogeneous morphism, φ_L. $\square$

Recall from Theorem 14.1 that for each cdga morphism $g : \varphi_A : A \to A'$ there is a morphism $\ell : L' \to L$ of profree dgl's, and a homotopy commutative diagram

$$
\begin{array}{ccc}
\wedge V_L & \xrightarrow[\psi]{\;\simeq\;} & A \\
\downarrow{\scriptstyle \varphi_\ell} & & \downarrow{\scriptstyle g} \\
\wedge V_{L'} & \xrightarrow[\psi']{\;\simeq\;} & A',
\end{array}
$$

in which ψ and ψ' are quasi-isomorphisms.

Proposition 14.2 *With the hypotheses and notation above, suppose also that L' is profree. Then:*

(i) *The correspondence $g \rightsquigarrow \varphi_\ell$ defines an inclusion of homotopy classes*

$$
i : [A, A'] \longrightarrow [L', L],
$$

associating with $g : A \to A'$ a profree dgl representative.

(ii) *ℓ is a homotopy equivalence $\iff$ ℓ is a quasi-isomorphism $\iff$ g is a quasi-isomorphism.*

(iii) *If $A = \wedge V$ is a Sullivan algebra, then the inclusion, i, is a bijection.*

Proof

(i) Given the diagram above we have only to show that the bijection $\ell \leftrightarrow \varphi_\ell$ induces a bijection $[L', L] \xrightarrow{\cong} [\wedge V_L, \wedge V_{L'}]$. Thus we only have to show that if $\varphi_0, \varphi_1 : \wedge V_L \to \wedge V_{L'}$ are bihomogeneous morphisms corresponding to profree dgl morphisms $\ell_0, \ell_1 : L \leftarrow L'$, then

$$\varphi_0 \sim \varphi_1 \iff \ell_0 \sim \ell_1.$$

Let $L' \,\widehat{\amalg}\, L' \,\widehat{\amalg}\, \overline{\mathbb{L}}_S$ be a cylinder object for L'. The inclusions $i_0, i_1 : L' \to L' \,\widehat{\amalg}\, L' \,\widehat{\amalg}\, \overline{\mathbb{L}}_S$ and the quasi-isomorphism $L' \,\widehat{\amalg}\, L' \,\widehat{\amalg}\, \overline{\mathbb{L}}_S \to L'$ correspond to bihomogeneous morphisms

$$\wedge V_{L'} \xrightarrow{\;\;\;\;\lambda\;\;\;\;} \wedge R \xrightarrow{\;\;\;\rho=(\rho_0,\rho_1)\;\;\;} \wedge V_{L'} \otimes \wedge V_{L'}$$

of semi-quadratic Sullivan algebras. Here $\wedge V_{L'}$ is a retract of $\wedge R$, ρ is surjective, and, since L' is profree, it follows from Proposition 12.2 that λ is a quasi-isomorphism. In particular, $\wedge R$ is a path object for $\wedge V_{L'}$.

Now if $\ell_0 \sim \ell_1 : L' \to L$, the homotopy $L' \,\widehat{\amalg}\, L' \,\widehat{\amalg}\, \overline{\mathbb{L}}_S \to L$ corresponds to a bihomogeneous morphism $\varphi : \wedge V_L \to \wedge R$. It is immediate from the definitions that φ is a homotopy from φ_0 to φ_1. On the other hand, suppose $\varphi : \wedge V_L \to \wedge R$ is a (not necessarily bihomogeneous) homotopy from φ_0 to φ_1. Since φ_0 and φ_1 are bihomogeneous, it is then straightforward to check that a bihomogeneous homotopy $\widehat{\varphi} : \wedge V_L \to \wedge R$ defined by

$$\widehat{\varphi} : V \to R \quad \text{and} \quad \varphi - \widehat{\varphi} : V \to \wedge^{\geq 2} R$$

corresponds to a homotopy $\ell_0 \sim \ell_1$.

(ii) This is immediate from Proposition 12.2 and Proposition 15.3.

(iii) In this case $[A, A'] = [\wedge V, \wedge V']$. $\square$

Theorem 14.2 *The cochain construction $L \mapsto \widehat{C}(L)$ induces a contravariant Quillen equivalence of model categories*

$$\text{Enrich} \longrightarrow \text{Cdga}$$

Proof Clearly if ℓ is a fibration, respectively, a cofibration, then $\widehat{C}(\ell)$ is a cofibration, respectively, a fibration. By Proposition 12.2, if ℓ is a weak equivalence, so is $\widehat{C}(\ell)$. By Theorem 14.1, $\widehat{C}$ induces a bijection between the weak equivalences classes of objects, and Proposition 14.2 implies a bijection of homotopy classes of maps in the derived category

$$[L, L'] = [\widehat{C}(L'), \widehat{C}(L).$$

$\square$

14.2 The Profree dgl Model of a Connected Space

To begin, we specialize the definitions and results of Part III as follows:

Definition

(i) A *profree dgl model of a connected space*, X, is a profree dgl, $(L(X), \partial)$, together
with a connecting quasi-isomorphism

$$(\wedge V_{L(X)}, d_0 + d_1) \xrightarrow{\;\cong\;} A_{PL}(X).$$

(ii) A *profree dgl representative of a continuous map*, $f : X \to Y$, of connected spaces
is a morphism $g : (L(X), \partial) \to (L(Y), \partial)$ of profree dgl models associated with a
homotopy commutative diagram of cdga's

$$
\begin{array}{ccc}
(\wedge V_{L(X)}, d_0 + d_1) & \xrightarrow{\;\simeq\;} & A_{PL}(X) \\[4pt]
\wedge_g \big\uparrow & & \big\uparrow A_{PL}(f) \\[4pt]
(\wedge V_{L(Y)}, d_0 + d_1) & \xrightarrow{\;\simeq\;} & A_{PL}(Y),
\end{array}
$$

in which $\wedge_g$ corresponds to g.

Remarks

1. Each connected space has a profree dgl model, and each continuous map between
   connected spaces has a profree dgl representative (Theorem 14.1).
2. If $f : X \to Y$ is a continuous map between connected spaces, then $H(f)$ is an
   isomorphism if and only if $\ell_f : (L(X), \partial) \to (L(Y), \partial)$ is a homotopy equivalence
   (Proposition 12.2). In particular, the profree dgl models of X form a single class of
   homotopy equivalent profree dgl's.

Proposition 14.3 *Let X and Y be connected spaces with minimal Sullivan models $\wedge W_X$
and $\wedge W_Y$ and with minimal profree dgl models $L(X)$ and $L(Y)$. Then we have bijections*

$$[X, Y_{\mathbb{Q}}] \cong [\wedge W_Y, \wedge W_X] \cong [L(X), L(Y)].$$

Proof The quasi-isomorphism $\wedge W_X \xrightarrow{\;\cong\;} A_{PL}(X)$ induces a bijection

$$[\wedge W_Y, \wedge W_X] \cong [\wedge W_Y, A_{PL}(X)].$$

On the other hand (7.3)

$$[\wedge W_Y, A_{PL}(X)] = [X, \langle \wedge W_Y \rangle] \cong [X, Y_{\mathbb{Q}}].$$

Finally from Proposition 14.2 we have

$$[\wedge V_Y, \wedge V_X] \cong [L(X), L(Y)].$$

$\square$

14.3   The Hurewicz Homomorphism

Proposition 14.4 *Let* $(\overline{\mathbb{L}}_T, \partial)$ *be a profree dgl model of a connected space* X *with finite type homology. Then there is a commutative diagram in which* $\rho : \overline{\mathbb{L}}_T \to T$ *denotes the projection with kernel* $\overline{\mathbb{L}}_T^{(2)}$.

$$
\begin{array}{ccccc}
\pi_*(X) & \longrightarrow & \pi_*(X_{\mathbb{Q}}) & \overset{\cong}{\longrightarrow} & sH(\overline{\mathbb{L}}_T, \partial) \\
\downarrow{\scriptstyle hur} & & \downarrow & & \downarrow{\scriptstyle sH(\rho)} \\
H_*(X; \mathbb{Z}) & \longrightarrow & H(X_{\mathbb{Q}})^{\vee} & \overset{\cong}{\longrightarrow} & sH(T, \overline{\partial}).
\end{array}
$$

This identifies $sH(\rho)$ *as the* Sullivan rationalization *of the Hurewicz homomorphism.*

Proof Let $(\wedge V, d_0 + d_1)$ be the semi-quadratic model associated with $\overline{\mathbb{L}}_T$ and $\wedge W_X$ its minimal Sullivan model. Then, Proposition 8.3 provides the commutative diagram

$$
\begin{array}{ccccc}
\pi_*(X) & \longrightarrow & \pi_*(X_{\mathbb{Q}}) & \overset{\cong}{\longrightarrow} & W_X^{\vee} \\
\downarrow{\scriptstyle hur} & & \downarrow & & \downarrow{\scriptstyle H(\xi)^{\vee}} \\
H_*(X; \mathbb{Z}) & \longrightarrow & H(X)^{\vee} & \overset{\cong}{\longrightarrow} & H(\wedge W_X)^{\vee},
\end{array}
$$

where $\xi : \wedge^{\geq 1} W_X \to W_X$ is division by $\wedge^{\geq 2} W_X$. Moreover (11.2)

$$H(V, d_0) \cong W_X \quad \text{and} \quad H(V \cap \ker d_1), d_0) \cong H(\wedge W_X).$$

Recall now that by Proposition 14.1 the correspondence between profree dgl and semi-quadratic Sullivan algebra gives the commutative diagram

$$s(\overline{\mathbb{L}}_T, \partial) \xrightarrow{\;\cong\;} (V, d_0)^{\vee}$$

with vertical maps $s\rho$ on the left and $j^{\vee}$ on the right, to

$$s(T, \bar{\delta}) \xrightarrow{\;\cong\;} (V \cap \ker d_1, d_0)^{\vee}.$$

This implies the result. $\square$

Proposition 14.5 *Let X be a simply connected space whose homotopy Lie algebra $L = L_X$ is profree. Then the Hurewicz map induces an injective map*

$$hur : L^{(2)}/L^2 \to H_*(X_{\mathbb{Q}}).$$

Proof Let α be an element in $L^{(2)}_{q-1}$, and write $\alpha = [g]$, with $g : S^q \to X_{\mathbb{Q}}$. Suppose $hur(\alpha) = 0$. Then there is a simply connected finite CW complex K of dimension $< q$ and a map $f : K \to X_{\mathbb{Q}}$ such that α is in the image of $f_* : \pi_q(K) \to \pi_*(X_{\mathbb{Q}})$ [6].

Denote by $\varphi : (L_K, \partial) \to (L_X, 0)$ a dgl representative of f, with (L_K, ∂) the minimal profree model of K. Thus $L_K = \widehat{\mathbb{L}}(W)$ with $W = W^{<q-1}$. Then let ω be a cycle in L_K^{n-1} representing α. Since all the elements in L_X are cycles, $\varphi(W)$ is a vector space of cycles. On the other hand by construction there is a cycle $\beta \in (L_K)_{q-1}$ such that $\varphi(\beta)$ is homologous to ω, thus equal to ω. But for degree reasons β is a decomposable element in L_K and so $\omega \in L^2$. This gives the result. $\square$

14.4   Wedges of Spheres and Disks

Here we assemble properties of a wedge of spheres, $X = \vee_{\alpha} S_{\alpha}^{n_{\alpha}}$, developed in Proposition 8.6.

(i) The homotopy Lie algebra of X is a profree dgl, $\overline{\mathbb{L}}_T$, and $(\overline{\mathbb{L}}_T, 0)$ is the profree dgl model of X.

(ii) The minimal Sullivan model of X is a quadratic Sullivan algebra $(\wedge V_S, d_1)$, for which there is a surjective quasi-isomorphism

$$(\wedge V_S, d_1) \xrightarrow{\;\cong\;} (\mathbb{Q} \oplus S, 0), \quad \text{with } S \cdot S = 0.$$

(iii) $H^{\geq 1}(X)^{\vee} \cong sT$.

On the other hand, the wedge $\vee_{\alpha} D^{n_{\alpha}+1}$ of disks is contractible. Thus the zero Lie algebra is a profree dgl model of $\vee_{\alpha} D^{n_{\alpha}+1}$. Therefore if $(\overline{\mathbb{L}}_T, 0)$ is a profree dgl model of $\vee_{\alpha} S^{n_{\alpha}}$, then

$$(\overline{\mathbb{L}}_T, 0) \to (\overline{\mathbb{L}}_T \,\hat{\sqcup}\, \overline{\mathbb{L}}_{sT}, \partial), \qquad \partial sx = x$$

is a profree dgl extension and a profree dgl representative of the inclusion $i : \vee_\alpha S^{n_\alpha} \to \vee_\alpha D^{n_\alpha+1}$.

Example (Product and Polyhedral Product of Circles) Let $X = (S^1)^n$. Since the minimal Sullivan model of X is $(\wedge(x_1, \ldots, x_n), 0)$ with each x_i in degree 1, its minimal profree dgl model L_X is given by the Tanré formula [78, (I.4)]:

$$L_X = (\overline{\mathbb{L}}_{(u_\tau)}, d),$$

where τ runs over all the non-empty subsets of $\{1, \ldots, n\}$. Here degree of $u_\tau = \mathrm{card}(\tau) - 1$, and

$$du_\tau = \sum_{q=1}^{card(\tau)} (-1)^q \sum_{(\sigma_1, \sigma_2)} \varepsilon(\sigma_1, \sigma_2)[u_{\sigma_1}, u_{\sigma_2}],$$

where (σ_1, σ_2) runs over all the $(q, \mathrm{card}\,\tau - q)$-shuffles and $\varepsilon(\sigma_1, \sigma_2)$ is the graded sign for the shuffle.

Now let $K \subset \{1, \ldots, n\}$ be a simplicial complex. The corresponding polyhedral product of circles $(S^1)^K$ is the subset of $(S^1)^n$ formed by the products $\prod_{i=1}^n X_i$ with $X_i = S^1$ or $X_i = *$ and such that the set of indices i with $X_i = S^1$ belongs to K. The minimal profree dgl model is then the corresponding sub dgl of $L_{(S^1)^n}$

$$(\overline{\mathbb{L}}_{(u_\sigma, \sigma \in K)}, d).$$

A Sullivan model for $(S^1)^K$ is given in [30]. For a simplex $\sigma \in \{1, \ldots, n\}$ we denote by $(\wedge x)^\sigma$ the tensor product $A_1 \otimes \cdots \otimes A_n$ where $A_i = *$ when $i \notin \sigma$ and $A_i = \wedge x_i$ otherwise. Then a model of $(S^1)^K$ is a model of

$$(\wedge x)^K = \wedge x_1 \otimes \cdots \otimes \wedge x_n / I_K$$

where I_K is the ideal generated by the $(\wedge x)^\sigma$ with σ not in K.

Now we consider a particular case. Let K_N be the union of all the simplices of dimension $\leq N$ in $\{1, \ldots, n\}$. Then I_{K_N} is the ideal $\wedge^N(x_1, \ldots, x_n)$. A tensor product with the acyclic closure of $\wedge(x_1, \ldots, x_n)$ gives the short exact sequence

$$0 \to I_{K_N} \otimes \wedge \overline{x}_i \to \wedge(x_i, \overline{x}_i) \to \left(\wedge(x_i)/\wedge^{>N} x_i\right) \otimes \wedge \overline{x}_i \to 0.$$

Since $H(I_{K_N} \otimes \wedge \overline{x}_i) = \mathbb{Q} \oplus H^N(I_{K_N} \otimes \wedge \overline{x}_i)$, the cohomology of $\wedge(x_i)/\wedge^{>N} x_i \otimes \wedge \overline{x}_i$ is concentrated in degree $N - 1$. Remark now that the minimal model of $(\wedge(x_i)/\wedge^{>N} x_i) \otimes \wedge(\overline{x}_i)$ is the minimal model of the homotopy fiber of the natural injection $(S^1)^K \to (S^1)^n$. Since the product of representing cocycles is zero, the homotopy fiber has the rational homotopy type of a wedge of spheres, and we have a short exact sequence

$$0 \to \overline{\mathbb{L}}_W \to \pi_* \Omega((S^1)^{K_N}) \otimes \mathbb{Q} \to Ab(x_1, \ldots, x_n),$$

where Ab denotes abelian Lie algebra and W is a vector space concentrated in degree $N - 2$.

14.5 Wedge of Spaces

Let more generally X and Y be path connected spaces with profree models $(\overline{\mathbb{L}}_V, \partial)$ and $(\overline{\mathbb{L}}_W, \partial)$. Then a profree model for the wedge $X \vee Y$ is the free product

$$(\overline{\mathbb{L}}_V, \partial) \,\widehat{\amalg}\, (\overline{\mathbb{L}}_W, \partial).$$

In particular we have isomorphisms of graded Lie algebras

$$s^{-1}\pi_*((X \vee Y)_{\mathbb{Q}}) \cong H_*((\overline{\mathbb{L}}_V \,\widehat{\amalg}\, \overline{\mathbb{L}}_W) \cong H_*(\overline{\mathbb{L}}_V, \partial) \,\widehat{\amalg}\, H_*(\overline{\mathbb{L}}_W, \partial)$$

$$\cong s^{-1}\pi_*(X_{\mathbb{Q}}) \,\widehat{\amalg}\, s^{-1}\pi_*(Y_{\mathbb{Q}}).$$

14.6 Cofibers

Proposition 14.6

(i) Let $L \xrightarrow{\varphi} (L \,\widehat{\amalg}\, \overline{\mathbb{L}}_Z, \partial) \xrightarrow{\rho} (\overline{\mathbb{L}}_Z, \overline{\partial})$ be a profree dgl extension, and let

$$\wedge V_L \xleftarrow{\psi} \wedge W \leftarrow \wedge V$$

be the diagram of corresponding semi-quadratic Sullivan algebras. Then there is a quasi-isomorphism $\wedge V \xrightarrow{\simeq} \mathbb{Q} \oplus \ker \psi$.

(ii) If φ is a dgl model for a map $f : X \to Y$, then $\rho : L \,\widehat{\amalg}\, \overline{\mathbb{L}}_Z \longrightarrow \overline{\mathbb{L}}_Z$ is a dgl model for the injection of Y in the homotopy cofiber of f.

Proof

(i) Let $\theta : (\overline{\mathbb{L}}_S, \partial) \xrightarrow{\simeq} L$ be a surjective profree dgl model of L. Then the decomposition of $\varphi \circ \theta$ as a profree extension followed by a surjective quasi-isomorphism (Proposition 13.3) gives the commutative diagram

$$
\begin{array}{ccccc}
(L, \partial) & \xrightarrow{\ \varphi\ } & (L \,\widehat{\amalg}\, \overline{\mathbb{L}}_Z, \partial) & \longrightarrow & (\overline{\mathbb{L}}_Z, \overline{\partial}) \\
\simeq \big\uparrow{\scriptstyle\theta} & & \simeq \big\uparrow{\scriptstyle\rho} & & \big\uparrow{\scriptstyle\rho'} \\
(\overline{\mathbb{L}}_S, \partial) & \xrightarrow{\ \varphi'\ } & (\overline{\mathbb{L}}_S \,\widehat{\amalg}\, \overline{\mathbb{L}}_{Z'}, \partial) & \longrightarrow & (\overline{\mathbb{L}}_{Z'}, \overline{\partial}).
\end{array}
$$

Now a direct application of Proposition 13.6 shows that the induced map ρ' : $\overline{\mathbb{L}}_{Z'} \to \overline{\mathbb{L}}_Z$ is a quasi-isomorphism.

Denote by

$$\wedge V_S \xleftarrow{\psi'} \wedge W' \xleftarrow{\eta} \wedge V' \text{ the semi-quadratic model of}$$

$$(\overline{\mathbb{L}}_S, \partial) \xrightarrow{\varphi'} (\overline{\mathbb{L}}_S \,\widehat{\amalg}\, \overline{\mathbb{L}}_{Z'}, \partial) \to (\overline{\mathbb{L}}_{Z'}, \overline{\partial}) \ .$$

We then have the commutative diagram

$$
\begin{array}{ccccc}
\wedge V_L & \xleftarrow{\ \psi\ } & \wedge W & \longleftarrow & \wedge V \\
\downarrow{\simeq} & & \downarrow{\simeq} & & \downarrow{\simeq} \\
\wedge V_S & \xleftarrow{\ \psi'\ } & \wedge W' & \longleftarrow & \wedge V'
\end{array}
$$

Since the vertical maps are injective, the induced map $\ker \psi \to \ker \psi'$ is a quasi-isomorphism. It is therefore enough to prove that $\wedge V'$ is quasi-isomorphic to $\mathbb{Q} \oplus \ker \psi'$.

Proposition 12.1 then provides the commutative diagram

$$
\begin{array}{ccccc}
s(S, \overline{\partial}) & \longrightarrow & s(S \oplus Z', \overline{\partial}) & \longrightarrow & s(Z', \overline{\partial}) \\
\downarrow{\simeq} & & \downarrow{\simeq} & & \downarrow{\simeq} \\
(\wedge^{\geq 1} V_S)^{\vee} & \longrightarrow & (\wedge^{\geq 1} W)^{\vee} & \longrightarrow & (\wedge^{\geq 1} V')^{\vee}.
\end{array}
$$

A final long exact homology sequence argument then shows that

$$(\ker \psi')^{\vee} = (\wedge^{\geq 1} W')^{\vee}/(\wedge^{\geq 1} V_S)^{\vee} \xrightarrow{\ \simeq\ } (\wedge^{\geq 1} V')^{\vee}.$$

It follows that the induced map $\wedge V' \to \mathbb{Q} \oplus \ker \psi'$ is a quasi-isomorphism.

(ii) By [33, Proposition 13.5] a Sullivan model for the injection of Y in the homotopy cofiber is given by the injection $\mathbb{Q} \oplus \ker \psi \to \wedge W$. This implies the result. $\qquad\square$

Now consider a commutative diagram of profree dgl extensions

$$
\begin{array}{ccc}
L & \xrightarrow{\ \ell_1\ } & L \,\widehat{\amalg}\, \overline{\mathbb{L}}_{T(1)} \\
\downarrow{\ell_0} & & \downarrow{\ell_2} \\
L \,\widehat{\amalg}\, \overline{\mathbb{L}}_{T(0)} & \xrightarrow{\ \ell_3\ } & L \,\widehat{\amalg}\, \overline{\mathbb{L}}_{T(1)\oplus T(0)}
\end{array} \ .
$$

with the corresponding diagram of semi-quadratic Sullivan algebras

$$\wedge W \xrightarrow{\;g_3\;} \wedge V(0)$$

with $g_2 : \wedge W \to \wedge V(1)$, $g_0 : \wedge V(0) \to \wedge V$, and $g_1 : \wedge V(1) \to \wedge V$.

each g_i denoting the semi-quadratic model of the corresponding ℓ_i.

Proposition 14.7

(i) With the above notation $\wedge W$ is quasi-isomorphic to $\wedge V(0) \times_{\wedge V} \wedge V(1)$.

(ii) If ℓ_0 and ℓ_1 are dgl models of the cofibrations $f_0 : X \to Y_0$ and $f_1 : X \to Y_1$, then $L \,\widehat{\amalg}\, \overline{\mathbb{L}}_{T(1)\oplus T(0)}$ is a dgl model for the pushout $Y_0 \cup_X Y_1$.

Proof

(i) Let (A, d) be the cdga pullback of the diagram of semi-quadratic Sullivan algebras; let $\theta : \wedge W \to A = \wedge V(1) \times_{\wedge V} \wedge V(0)$ be the induced map

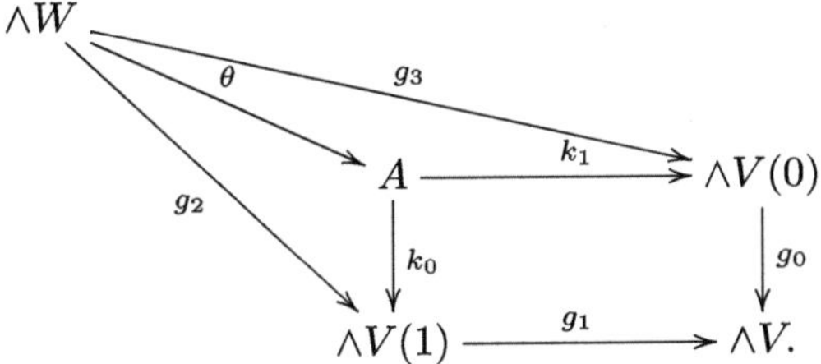

By Proposition 14.6, the cdga's $\mathbb{Q} \oplus \ker g_0$ and of $\mathbb{Q} \oplus \ker g_1$ are quasi-isomorphic to the Sullivan model of the quotient dgl's $(\overline{\mathbb{L}}_{T(0)}, \overline{\partial})$ and $(\overline{\mathbb{L}}_{T(1)}, \overline{\partial})$. On the other hand by definition of pullback the kernel of g_0 is isomorphic to the kernel of k_0. Since $g_2 : \wedge W \to \wedge V(1)$ is a Sullivan model of the dgl extension

$$L \,\widehat{\amalg}\, \overline{\mathbb{L}}_{T(1)} \to L \,\widehat{\amalg}\, \overline{\mathbb{L}}_{T(1)} \,\widehat{\amalg}\, \overline{\mathbb{L}}_{T(0)},$$

$\ker g_0 \simeq \ker g_2$.

Finally the commutativity of the diagram

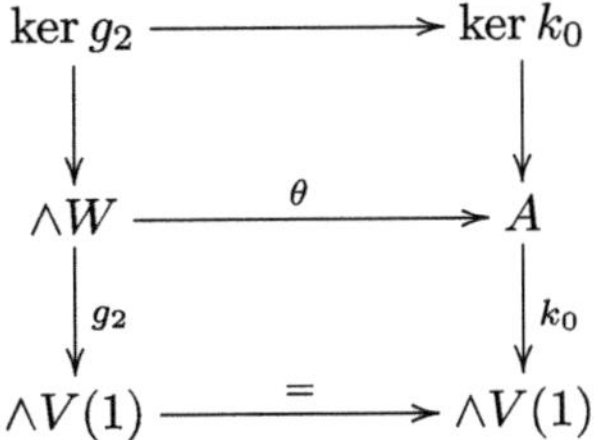

implies that θ is a quasi-isomorphism.

(ii) This follows directly from [33, Proposition 13.5]. $\qquad\square$

14.7   The Quotient Space, X/Y

Let $f : X \to Y$ be a continuous map of connected spaces, but assume also that f is the inclusion of a sub-simplicial set, or of a sub-CW complex, or of a deformation retract of an open set

$$X \xrightarrow{\quad f \quad} Y \xrightarrow{\quad \pi \quad} Y/X.$$

Now, denote by $L(Y)$ and $L(X)$ profree dgl models of Y and X. A dgl representative of f is then a morphism $\ell_f : L(X) \to L(Y)$ that factors as

$$\ell_f : L(X) \xrightarrow{\quad \lambda \quad} L(X) \,\widehat{\amalg}\, \overline{\mathbb{L}}_T \xrightarrow[\simeq]{\quad \gamma \quad} L(Y),$$

in which λ is a profree dgl extension and γ is a quasi-isomorphism.

Proposition 14.8 *With the hypotheses above, the surjection*

$$\rho : L(Y) \,\widehat{\amalg}\, \overline{\mathbb{L}}_T \longrightarrow \overline{\mathbb{L}}_T$$

is a profree dgl representative of the map $\pi : Y \to Y/X$. In particular, $\overline{\mathbb{L}}_T$ is a profree dgl model of Y/X.

Proof This is a particular case of Proposition 14.6. $\qquad\square$

Part IV

Cell Attachments

This section is devoted to cell attachments and more precisely to their rationalizations. We exhibit models for the attachments in terms of minimal Sullivan models and in terms of profree dgl models. This section ends with the example of a connected finite CW complex whose homotopy Lie algebra is not finitely generated.

Definition A (*topological*) *attaching map* is a continuous map

$$g : \vee_\alpha S^{n_\alpha} \to X$$

from a wedge of connected spheres to a connected space, X. Then the map

$$\iota_X : X \to X \cup_g \vee_\alpha D^{n_\alpha+1}$$

is a (*topological*) *cell attachment.*

Given such a cell attachment, observe that $X \cup_g \vee_\alpha D^{n_\alpha+1}$ is the pushout of the diagram

$$
\begin{array}{ccc}
\vee_\alpha S^{n_\alpha} & \xrightarrow{\ \ g\ \ } & X \\
\downarrow{\scriptstyle i} & & \downarrow{\scriptstyle i_X} \\
\vee_\alpha D^{n_\alpha+1} & \longrightarrow & X \cup_g \vee_\alpha D^{n_\alpha+1}.
\end{array}
$$

Definition

(i) A *dgl attaching map* is a morphism

© The Author(s), under exclusive license to Springer Nature Switzerland AG 2026

Y. Félix, S. Halperin, *Lie Models for Spaces*, Frontiers in Mathematics,

https://doi.org/10.1007/978-3-032-15357-9_15

$$\gamma : (\overline{\mathbb{L}}_T, 0) \to (L, \partial_L)$$

of enriched dgl's, and the corresponding dgl cell attachment is the inclusion

$$j : L \to L \,\widehat{\amalg}\, \overline{\mathbb{L}}_{sT}, \qquad \partial sx = \gamma(x).$$

(ii) The *homology attaching map associated with* γ is the dgl attaching map

$$H(\gamma) : (\overline{\mathbb{L}}_T, 0) \to (H(L), 0).$$

Proposition 15.1 *Let* $g \,:\, \vee_\alpha S^{n_\alpha} \,\to\, X$ *be a topological attaching map and* $\gamma \,:\, (\overline{\mathbb{L}}_T, 0) \longrightarrow (L, \partial)$ *a profree dgl representative of g. Then the injection* $j \,:\, (L, \partial) \to (L \,\widehat{\amalg}\, \overline{\mathbb{L}}_{sT}, \partial)$, $\partial sx = \gamma(x) \in L$, *is a profree dgl representative of* i_X.

Proof Let θ be a minimal dgl cofibration model of γ. This gives the commutative diagram

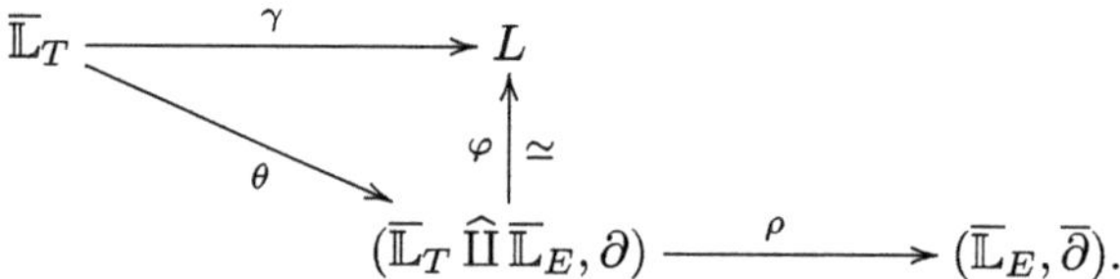

By Proposition 14.6, ρ is a dgl model for i_X. We then have the commutative diagram

$$
\begin{array}{ccc}
L & \xrightarrow{\quad j \quad} & (L \,\widehat{\amalg}\, \overline{\mathbb{L}}_{sT}, \partial) \\
\varphi \uparrow\, \simeq & & \simeq \,\uparrow\, \varphi\,\widehat{\amalg}\, id \\
(\overline{\mathbb{L}}_T \,\widehat{\amalg}\, \overline{\mathbb{L}}_E, \partial) & \longrightarrow & (\overline{\mathbb{L}}_T \,\widehat{\amalg}\, \overline{\mathbb{L}}_E \,\widehat{\amalg}\, \overline{\mathbb{L}}_{sT}, \partial') \xrightarrow{\ \simeq\ } (\overline{\mathbb{L}}_E, \overline{\partial}), \partial'(st) = t.
\end{array}
$$

$\rho :$ (at left)

It follows that j is also a dgl model for i_X. $\qquad\qquad\qquad\qquad\qquad\qquad\qquad\square$

Example *Attachment of 2-cells to a finite wedge of circles*

Suppose $g \,:\, Y = \vee_{i=1}^{r} S^1_{a_i} \to X = \vee_{j=1}^{s} S^1_{b_j}$ is an attaching map with corresponding cell attachment

$$i_X : X \to X \cup_g \vee_i D^2_{a_i}.$$

Then g determines and is determined by a group homomorphism

$$\omega : G_a \to G_b$$

between the corresponding fundamental groups. Both groups are free and freely generated, respectively, by element $a_1, \ldots, a_r$ and $b_1, \ldots, b_s$ corresponding to the circles.

Let $L(g) : (L(Y), 0) \to (L(X), 0)$ be a profree dgl model of g. By (21), we have the commutative diagram

$$
\begin{array}{ccc}
sL(Y) & \xrightarrow{\ sL(g)\ } & sL(X) \\
\Big\downarrow{\scriptstyle =} & & \Big\downarrow{\scriptstyle =} \\
V_Y^\vee & \xrightarrow[\ \psi^\vee\ ]{} & V_X^\vee,
\end{array}
$$

where $\psi : (\wedge V_Y, d_1) \longleftarrow (\wedge V_X, d_1)$ is a minimal Sullivan representative of g. On the other hand, by Proposition 8.12,

$$
L_Y = \overline{\mathbb{L}}_{T(a)} \quad \text{and } L_X = \overline{\mathbb{L}}_{T(b)}
$$

in which $T(a)$ has as a basis the elements $\log_{G_a}(a_i)$ and $T(b)$ has as a basis the elements $\log_{G_b}(b_j)$. Moreover, it is immediate from the definitions (Sect. 9.3) that the diagram

$$
\begin{array}{ccccc}
G_a & \longrightarrow & G_{L_Y} & \xrightarrow{\ \log\ } & sL_Y \\
\Big\downarrow{\scriptstyle \omega} & & \Big\downarrow & & \Big\downarrow{\scriptstyle s\ell_g} \\
G_b & \longrightarrow & G_{L_X} & \xrightarrow[\ \log\ ]{} & sL_X
\end{array}
$$

commutes. Therefore

$$
\ell_g(\log_{G_a}(a_i)) = \log_{G_b}(\omega a_i).
$$

Thus it follows from Proposition 15.1 that a dgl representative of i_X is the morphism

$$
L_X \to (L_X \,\widehat{\amalg}\, \overline{\mathbb{L}}_{sT(a)}, \partial) \quad \text{with } \partial(s \log_{G_a} a_i) = \log_{G_b}(\omega a_i).
$$

15.1   dgl Attaching Maps and Sullivan Attaching Maps

Let $\gamma : (\overline{\mathbb{L}}_T, 0) \to (L, \partial_L)$ be a dgl attaching map. The corresponding bihomogeneous morphism of semi-quadratic Sullivan algebras then has the form (Sect. 16.2)

$$
\gamma' : (\wedge V_L, d_0 + d_1) \to (\wedge V_S, d_1)
$$

in which $(\wedge V_S, d_1)$ is a quadratic Sullivan algebra, $\overline{\mathbb{L}}_T$ is the homotopy Lie algebra of $(\wedge V_S, d_1)$, and $q : (\wedge V_S, d_1) \xrightarrow{\ \simeq\ } (\mathbb{Q} \oplus S, 0)$ with $S \cdot S = 0$.

Now let $(\wedge W, d)$ be the minimal Sullivan model of $(\wedge V_L, d_0 + d_1)$, so that we have a surjective quasi-isomorphism $\varphi : (\wedge V_L, d_0 + d_1) \xrightarrow{\simeq} (\wedge W, d)$. Since $(\wedge W, d)$ is minimal, φ admits a retract σ. We define $\rho : \wedge W \to \mathbb{Q} \oplus S$ as the composition $\rho = q j' \sigma$. There results the homotopy commutative diagram

$$
\begin{array}{ccc}
(\wedge V_L, d_0 + d_1) & \xrightarrow[\simeq]{\varphi} & (\wedge W, d) \\
\Big\downarrow{\gamma'} & & \Big\downarrow{\rho} \\
(\wedge V_S, d_1) & \xrightarrow[\simeq]{q} & (\mathbb{Q} \oplus S, 0).
\end{array}
$$

Definition

(i) A *Sullivan attaching map* is a morphism

$$\rho : (\wedge W, d) \to (\mathbb{Q} \oplus S, 0)$$

from a minimal Sullivan algebra in which $S \cdot S = 0$.

(ii) The *quadratic attaching map* associated with ρ is the morphism

$$\rho(1) : (\wedge W, d_1) \to (\mathbb{Q} \oplus S, 0)$$

defined by $\rho(1)w = \rho(w)$.

Proposition 15.2

(i) The correspondence $[\gamma] \mapsto [\rho]$ induces a bijection between sets of homotopy classes

$$[\overline{\mathbb{L}}_T, L] \xrightarrow{\cong} [\wedge W, \mathbb{Q} \oplus S].$$

(ii) $H(\gamma) : T \to H(L)$ is injective if and only if $\rho : \wedge W \to \mathbb{Q} \oplus S$ is surjective.

Proof

(i) is a particular case of Proposition 14.2.
(ii) This follows immediately from the contravariant bijection

$$[T, H(L)] \xrightarrow{\cong} [W, S].$$

$\square$

15.2   A cdga Representative of a dgl Cell Attachment $j : L \to L \,\widehat{\amalg}\, \overline{\mathbb{L}}_{sT}$

We maintain the notation of the earlier sections. In particular, let $\gamma : \overline{\mathbb{L}}_T \to L$ be a dgl attaching map and $\rho : (\wedge W, d) \to (\mathbb{Q} \oplus S, 0)$ its associated Sullivan attaching map. Then ρ factors to define a linear map

$$\overline{\rho} : W \to S.$$

This in turn defines two cdga's:

- $(A, d) := (\mathbb{Q} \oplus \ker \overline{\rho}) \oplus \wedge^{\geq 2} W \subset \wedge W$.
- $(\mathbb{Q} \oplus s \operatorname{coker} \overline{\rho}, 0)$ with $s \operatorname{coker} \overline{\rho} \cdot s \operatorname{coker} \overline{\rho} = 0$.

The next proposition is a generalization of Proposition 14.6 corresponding to the situation where $\overline{\rho} : W \to S$ is not surjective.

Proposition 15.3 *Denote by* $\wedge V_R \longrightarrow \wedge W$ *the morphism of semi-quadratic Sullivan algebras corresponding to* $j : L \to L \,\widehat{\amalg}\, \overline{\mathbb{L}}_{sT}$. *Then there is a quasi-isomorphism*

$$\varphi : \wedge V_R \xrightarrow{\;\simeq\;} A \times_{\mathbb{Q}} (\mathbb{Q} \oplus s \operatorname{coker} \overline{\rho}).$$

Moreover the morphism $A \times_{\mathbb{Q}} (\mathbb{Q} \oplus s \operatorname{coker} \overline{\rho}) \to \wedge W$ *that is the injection on A and maps* $s \operatorname{coker} \overline{\rho}$ *to 0 is a cdga representative of* i_X.

Proof Let θ be a minimal dgl cofibration model of γ. This gives the commutative diagram of enriched dgl's

$$
\begin{array}{ccc}
\overline{\mathbb{L}}_T & \xrightarrow{\;\;\gamma\;\;} & L \\
 & {\scriptstyle \theta}\searrow & {\scriptstyle \varphi}\big\uparrow{\scriptstyle \simeq} \\
 & & (\overline{\mathbb{L}}_T \,\widehat{\amalg}\, \overline{\mathbb{L}}_E, \partial) \longrightarrow (\overline{\mathbb{L}}_E, \overline{\partial}),
\end{array}
$$

and the associated commutative diagram of semi-quadratic Sullivan algebras

$$
\begin{array}{ccccc}
\mathbb{Q} \oplus S & \xleftarrow{\;\simeq\;} \wedge Z & \longleftarrow & \wedge W & \longleftarrow \wedge V_S \\
 & & {\scriptstyle \psi}\searrow & \big\downarrow{\scriptstyle \simeq} & \\
 & & & \wedge V & \longleftarrow \wedge V_E.
\end{array}
$$

We then have the following commutative diagram:

$$
\begin{array}{ccc}
L & \xrightarrow{\;\;\;j\;\;\;} & (L\,\widehat{\amalg}\,\overline{\mathbb{L}}_{sT},\partial) \\[2pt]
\varphi \,\Big\uparrow\, \simeq & & \varphi\,\widehat{\amalg}\,id\,\Big\uparrow\,\simeq \\[2pt]
\rho: \quad (\overline{\mathbb{L}}_T\,\widehat{\amalg}\,\overline{\mathbb{L}}_E,\partial) & \longrightarrow & (\overline{\mathbb{L}}_T\,\widehat{\amalg}\,\overline{\mathbb{L}}_E\,\widehat{\amalg}\,\overline{\mathbb{L}}_{sT},\partial') \xrightarrow{\;\;\simeq\;\;} (\overline{\mathbb{L}}_E,\overline{\partial}), \quad \partial'(st)=t.
\end{array}
$$

Then, by Proposition 14.6, $\wedge V_E \xrightarrow{\;\simeq\;} \mathbb{Q}\oplus\ker\psi$. Now we form the cdga

$$
B = \mathbb{Q}\oplus\operatorname{coker}\overline{\rho}\oplus s\operatorname{coker}\overline{\rho},
$$

with trivial multiplication and $d(a)=sa$ for $a\in\operatorname{coker}\overline{\rho}$.

Finally let $\sigma:\operatorname{coker}\overline{\rho}\to S$ be a section of the projection. This gives the commutative diagram of surjections

$$
\begin{array}{ccc}
\mathbb{Q}\oplus S & \xleftarrow{\quad\rho\quad} & \wedge W \\[4pt]
& \searrow{\scriptstyle\rho'} & \Big\downarrow{\simeq} \\[4pt]
& & \wedge W \times_{\mathbb{Q}} B
\end{array}
$$

with $\rho'(a)=\sigma a$ for $a\in\operatorname{coker}\overline{\rho}$, and $\rho'(sa)=0$.
It follows that

$$
\mathbb{Q}\oplus\ker\rho \simeq \mathbb{Q}\oplus\ker\rho' = (\mathbb{Q}\oplus\ker\rho)\times_{\mathbb{Q}}(\mathbb{Q}\oplus s\operatorname{coker}\overline{\rho}).
$$

This gives the result.

The final statement is a direct consequence of the commutative diagram

$$
\begin{array}{ccc}
\wedge V_R & \xrightarrow{\hspace{5cm}} & \wedge W \\[4pt]
\Big\downarrow{\simeq} & & \Big\downarrow{\simeq} \\[4pt]
A\times_{\mathbb{Q}}(\mathbb{Q}\oplus s\operatorname{coker}\overline{\rho}) & \longrightarrow & \wedge W\times_{\mathbb{Q}}(\mathbb{Q}\oplus\operatorname{coker}\overline{\rho}\oplus s\operatorname{coker}\overline{\rho}.
\end{array}
$$

$\square$

Proposition 15.4 *With the notations above, suppose $\rho:\wedge W\to\mathbb{Q}\oplus S$ is surjective.*

(i) A Sullivan representative of $j:L\to L\,\widehat{\amalg}\,\overline{\mathbb{L}}_{sT}$ is the morphism $\lambda:\wedge V\to\wedge W$, where $\wedge V$ is the minimal model of $\mathbb{Q}\oplus\ker\rho$.

(ii) Let $T=s^{-1}S$, and consider the cdga $(\wedge W\oplus T, D)$ in which

$$
Dw = dw + s^{-1}\rho w
$$

and $W\cdot T=0$. Then the injection

$$\theta : A \to \wedge W \oplus T$$

is a cdga quasi-isomorphism.

Example Recall that the fundamental group of $S_a^1 \vee S_b^1$ is a free group on two generators, a, b. Denote by X the space obtained by adding two cells of dimension 2 to $S_a^1 \vee S_b^1$ to oblige a and b to commute with the commutator $aba^{-1}b^{-1}$. Then $\pi_1(X)$ is a metabelian group, and the Lie algebra associated with its Malcev completion is the quotient of the completion of the free Lie algebra on two elements a, b by the relations $[a, [a, b]] = [b, [a, b]] = 0$.

Now we prove that the Lie algebra $\pi_{\geq 2}(X_{\mathbb{Q}})$ is profree. Denote by $(\wedge V, d)$ the minimal model of X. The space V^1 has a basis with three elements x, y, t with $dx = dy = 0, dt = xy$, and $(\wedge V, d)$ is quasi-isomorphic to $\wedge (x, y, t)/xyt$. Decompose $\wedge V = \wedge V^1 \otimes \wedge Z$, and denote by $\wedge V^1 \otimes \wedge U$ the acyclic closure of $\wedge V^1$. The short exact sequence

$$0 \to \mathbb{Q}xyt \to \wedge(x, y, t) \to \wedge(x, y, t)/(xyt) \to 0$$

shows that

$$(\wedge Z, \overline{d}) \xrightarrow{\simeq} (\mathbb{Q} \oplus s^{-1}(xyt \otimes \wedge U), 0).$$

It follows that L_Z is a profree Lie algebra.

15.3   A Connected Finite CW Complex, Y, Whose Homotopy Lie Algebra, L_Y, Is Not Nilpotent and Satisfies $\dim (L_Y/L_Y^2)_1 = \infty$

Step 1. Construction of Y and a first profree dgl model for Y
In [59] Jean-Michel Lemaire has constructed a simply connected finite CW complex, X, whose homotopy Lie algebra, L_X, is not finitely generated. Since L_X is a graded vector space of finite type, the same is true for L_X/L_X^2, and so the generators occur in infinitely many degrees.

Lemaire's example is constructed by attaching seven cells to a wedge of five spheres. Here we use a similar process to construct a CW complex, Y by attaching seven 2-cells $D_{a_i}^2$ to a wedge of five circles $S_{b_j}^1$ via a map

$$g : \vee_{i=1}^7 S_{a_i}^1 \longrightarrow \vee_{j=1}^5 S_{b_j}^1.$$

The respective fundamental groups are free groups G_a and G_b freely generated by generators $a_1, \ldots, a_7$ and $b_1, \ldots, b_5$.

We denote the commutators in G_b by

$$(c, d) = cdc^{-1}d^{-1}.$$

Then g is defined by the homomorphism $\pi_1(g)$ given explicitly by

$$a_1 \mapsto (b_1, b_3),\, a_2 \mapsto (b_1, b_4),\, a_3 \mapsto (b_2, b_3),\, a_4 \mapsto (b_2, b_4), \quad \text{and}$$

$$a_5 \mapsto (b_5, b_1 b_3^{-1}),\, a_6 \mapsto (b_5, b_1 b_4^{-1}),\, a_7 \mapsto (b_5, b_2 b_3^{-1}).$$

Now set $X = \vee_{j=1}^{5} S_{b_j}^1$, and define

$$Y = X \cup_g \vee_{i=1}^{7} D_{a_i}^2.$$

As usual the attachment is denoted by $\iota_X : X \to X \cup_g (\vee_{i=1}^{7} D_{a_i}^2) = Y$.

We use Proposition 9.3 to construct a first profree dgl model of Y. As in Sect 9.1, denote by $\log_{G_b}$ the map

$$\log_{G_b} : G_b \longrightarrow G_{L_X} \xrightarrow{\ \log\ } L_X,$$

recalling that $\log$ is a bijection of sets. Then L_X is a profree Lie algebra $\overline{\mathbb{L}}_{T(b)}$ in which the elements $\log_{G_b}(b_j)$ are a basis of $T(b)$ (Proposition 9.4(i)). In the same way we denote by $(\overline{\mathbb{L}}_{T(a)}, 0)$ a dgl model for the wedge of the seven spheres $S_{a_i}^1$. Then a profree dgl representative of ι_X is given by

$$(L_X, 0) \to (L_X \,\widehat{\amalg}\, \overline{\mathbb{L}}_{sT(a)}, \partial),$$

in which the elements $\log_{G_a}(a_i)$ are a basis of $T(a)$ and

$$\partial \left(s \log_{G_a}(a_i) \right) = \log_{G_b}(\pi_1(g)(a_i)).$$

In particular, $(L_X \,\widehat{\amalg}\, \overline{\mathbb{L}}_{sT(a)}, \partial)$ is a profree dgl model for Y.

To simplify notation we set

$$\log_{G_b} b_j = x_j \quad \text{and} \quad \log_{G_a}(a_i) = y_i.$$

Then the dgl morphism above can be written as

$$(L_X, 0) \to L_X \,\widehat{\amalg}\, \overline{\mathbb{L}}_{sT(a)} = \overline{\mathbb{L}}_{(x_1,\dots x_5)} \,\widehat{\amalg}\, \overline{\mathbb{L}}_{(sy_1,\dots,sy_7)}$$

in which $\partial(sy_i) = \log_{G_b}(\pi_1(g)(a_i))$. Specifically,

$$\partial sy_1 = \log_{G_b}(b_1, b_3), \ \partial sy_2 = \log_{G_b}(b_1, b_4), \ \partial sy_3 = \log_{G_b}(b_2, b_3), \ \partial sy_4$$

$$= \log_{G_b}(b_2, b_4)$$

and

$$\partial sy_5 = \log_{G_b}(b_5, b_1 b_3^{-1}), \ \partial sy_6 = \log_{G_b}(b_5, b_1 b_4^{-1}), \quad \text{and } \partial sy_7 = \log_{G_b}(b_5, b_2 b_3^{-1}).$$

Step 2. A second profree dgl model of Y
We define a second profree dgl model $(L_X \,\widehat{\amalg}\, \overline{\mathbb{L}}(sy_1, \ldots, sy_7), \partial')$ by setting

$$\partial' sy_1 = [x_1, x_3], \partial' sy_2 = [x_1, x_4], \partial' sy_3 = [x_2, x_3], \partial' sy_4 = [x_2, x_4], \quad \text{and}$$

$$\partial' sy_5 = [x_5, x_1 - x_3], \partial' sy_6 = [x_5, x_1 - x_4] \quad \text{and } \partial' sy_7 = [x_5, x_2 - x_3].$$

We then identify this as a profree dgl model for Y by constructing a commutative diagram
of dgl morphisms

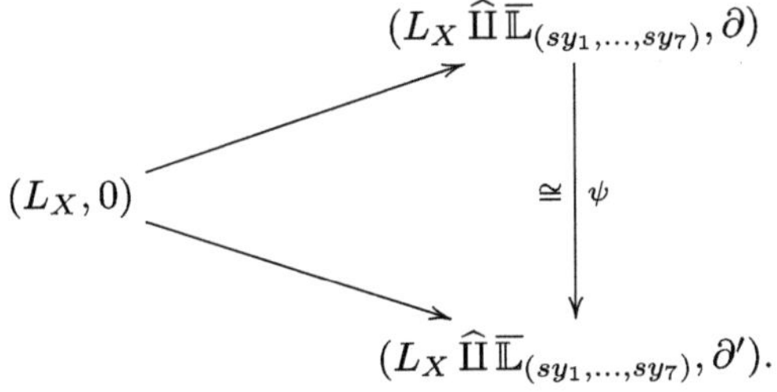

For this, denote by $\overline{J}$ the augmentation ideal of $\overline{UL_X}$. Then the left and right adjoint
representations of L_X extend to representations of $\overline{UL_X}$. In particular, if $u \in L_X$ and
$\alpha \in \overline{J}$, then

$$u \cdot \alpha = \alpha^t \cdot u,$$

where if $\alpha = \alpha_1 \ldots \alpha_k$ with $\alpha_i \in L_X$, then

$$\alpha^t = (-1)^k \alpha_k \ldots \alpha_1.$$

More generally, if $\alpha = \sum \alpha_k$, with $\alpha_k \in \overline{J}^k$, then $\alpha^t = \sum \alpha_k^t$.
   Similarly, if $\alpha = \alpha_1 \ldots \alpha_k \in \overline{J}$, with $\alpha_i \in L_X$, we define $\alpha \cdot sy_j \in L_X \,\widehat{\amalg}\, \overline{\mathbb{L}}(sy_j)$ by
setting

$$\alpha \cdot sy_j = [\alpha_1, [\alpha_2, \ldots [\alpha_k, sy_j] \ldots]].$$

It is then immediate that for any $u \in L_X$, $\alpha \in \overline{J}$, and $1 \leq i \leq 7$,

$$[u \cdot \alpha, sy_i] = [u, \alpha \cdot sy_i].$$

We now define ψ. It is immediate from Proposition 9.3(ii) that for some $\alpha_1, \ldots, \alpha_4 \in \overline{J}$,

$$\partial sy_1 = [x_1, x_3] + [x_1, x_3] \cdot \alpha_1, \qquad \partial sy_2 = [x_1, x_4] + [x_1, x_4] \cdot \alpha_2,$$

and

$$\partial sy_3 = [x_2, x_3] + [x_2, x_3] \cdot \alpha_3, \qquad \partial sy_4 = [x_2, x_4] + [x_2, x_4] \cdot \alpha_4.$$

We define

$$\psi(sy_i) = sy_i + \alpha_i \cdot sy_i, \qquad 1 \leq i \leq 4.$$

Next, observe that for some β and $\alpha_5 \in \overline{J}$

$$\log_{G_b}(b_1 b_3^{-1}) = x_1 - x_3 - \frac{1}{2}[x_1, x_3] + [x_1, x_3] \cdot \beta,$$

and

$$(x_5, \log_{G_b}(b_1 b_3^{-1})) = [x_5, \log_{G_b}(b_1 b_3^{-1})] + [x_5, \log_{G_b}(b_1, b_3^{-1})] \cdot \alpha_5.$$

It follows that

$$[x_5, \log_{G_b}(b_1 b_3^{-1})] = [x_5, x_1 - x_3] + \frac{1}{2}[x_1, x_3], x_5] - [x_1, x_3] \cdot \beta x_5.$$

This gives

$$\partial sy_5 = [x_5, x_1 - x_3] + [x_1, x_3] \cdot \beta_5' = [x_5, x_1 - x_3] + \beta_5 \cdot [x_1, x_3],$$

for some β_5' and β_5 in $\overline{J}$. Similarly,

$$\partial sy_6 = [x_5, x_1 - x_4] + \beta_6 \cdot [x_1, x_4] \quad \text{and} \quad \partial sy_7 = [x_5, x_2 - x_3] + \beta_7 \cdot [x_2, x_3].$$

Now we define

$$\psi(sy_5) = sy_5 + \beta_5 \cdot sy_1, \; \psi(sy_6) = sy_6 + \beta_6 sy_2, \; \psi(sy_7) = sy_7 + \beta_7 \cdot sy_3.$$

Then $\partial'\psi = \psi\partial$, and it is immediate that ψ is an isomorphism.

Step 3. $dim\,(L_Y/L_Y^2)_1 = \infty$

We first consider the uncompleted dgl $(L, d') = (\mathbb{L}(x_i, sy_j), d')$. Following the computations of [59, page 118, 119], the dimension of $(H(L, d')/H(L, d')^2)_1$ is infinite.

Denote by L_s^r the vector space generated by the Lie brackets containing exactly r elements sy_j and s elements x_i. Since $d(sy_j) \subset \mathbb{L}^2(x_i)$, the dgl L is the sum of the finite dimensional complexes $C_n = \oplus_{2r+s=n} L_s^r$. Moreover $\sum_{r\geq r_0} C_r$ is an ideal, and we denote by $L(r_0)$ the quotient dgl $L/(\sum_{r\geq r_0} C_r)$. Thus, for any integer k, there is an N such that $\dim\,(L(N)/L(N)^2)_1 > k$.

It follows that its completion $\overline{\mathbb{L}}_{(x_i,sy_j)}$ decomposes as the product $\prod_n C_n$, so that its homology is $\prod_n H_*(C_n)$, the completion of $H(L)$.

Therefore

$$L_Y = H(\overline{\mathbb{L}}_{(x_i,sy_j)}, d) \cong \prod_n H(C_n).$$

Remark now that $\prod_{r\geq r_0} H(C_r)$ is an ideal in L_Y whose quotient is $L(r_0)$. Therefore if $\dim\,(L_Y/L_Y^2)_1 \leq k$ for some integer k, then for any r_0, $\dim\,(L(r_0)/L(r_0)^2) \leq k$, which is not true. It follows that $\dim\,(L_Y/L_Y^2)_1 = \infty$.

Inert Attachments **16**

In topology a cell attachment $f : S^n \to X$ is inert if the attachment does not create any new homotopy class, more precisely, if the induced map $\pi_*(X) \to \pi_*(X \cup_f e^{n+1})$ is surjective. Here we consider the problem from the rational homotopy point of view and give criteria for $\pi_*(X_\mathbb{Q}) \to \pi_*((X \cup_f e^{n+1})_\mathbb{Q})$ to be surjective. Equivalent characterizations are given using minimal Sullivan models and profree dgl models.

16.1   dgl vs. Sullivan Attaching Maps

Recall that a topological attaching map and the induced cell attachment are continuous maps of connected spaces

$$g : \vee_\alpha S^{n_\alpha} \to X \quad \text{and} \quad i_X : X \to Y = X \cup_g \vee_\alpha D^{n_\alpha+1}.$$

While $\operatorname{Im} \pi_*(g) \subset \operatorname{Ker} \pi_*(\iota_X)$, this construction will often create additional elements in $\pi_*(Y)$, and it is far from understood which conditions will assure that

$$\pi_*(i_X) : \pi_*(X) \to \pi_*(Y)$$

is surjective.

This condition has been first considered in rational homotopy for simply connected spaces in [5] and [41]. Consistent with the simply connected case, we introduce the following definition:

Y. Félix, S. Halperin, *Lie Models for Spaces*, Frontiers in Mathematics,
https://doi.org/10.1007/978-3-032-15357-9_16

Definition The attaching map, g, is *rationally inert* if

$$\pi_*(i_X)_{\mathbb{Q}} : \pi_*(X_{\mathbb{Q}}) \to \pi_*(Y_{\mathbb{Q}})$$

is surjective.

In the case X is simply connected with finite type homology, the inertness of g is equivalent to the surjectivity of $\pi_*(i_X) \otimes \mathbb{Q}$, or saying that rationally the loop map Ωi_X has a right homotopy inverse. This condition has been generalized over the integers in [79], and the map g is called *inert* if the loop space $\Omega i_X : \Omega X \to \Omega Y$ has a right homotopy inverse.

Here we extend the concept of inert to the category of profree dgl's:

Definition A dgl attaching map $\gamma : (\overline{\mathbb{L}}_T, 0) \to (L, \partial_L)$ is *inert* if the induced map

$$H(j) : H(L) \to H(L \,\widehat{\amalg}\, \overline{\mathbb{L}}_{sT})$$

is surjective.

The image of T in $H(L)$ is then called an *inert subspace*.
We then have the following proposition:

Proposition 16.1 *Let γ and j be the profree dgl representatives of g and i_X*

$$\gamma : (\overline{\mathbb{L}}_T, 0) \to (L, \partial_L) \quad and \quad j : (L, \partial_L) \to (L \,\widehat{\amalg}\, \overline{\mathbb{L}}_{sT}, \partial).$$

Then, in particular,

$$g : \vee_\alpha S^{n_\alpha} \to X \text{ is rationally inert} \iff \text{a profree dgl representative } \gamma \text{ is inert.}$$

Lemma 16.1

(i) *If $\gamma \sim \gamma' : (\overline{\mathbb{L}}_T, 0) \to (L, \partial_L)$, then γ is inert if and only if γ' is inert.*
(ii) *If $\gamma : (\overline{\mathbb{L}}_T, 0) \to (L, \partial_L)$ is inert, then $H(\gamma) : T \to H(L)$ is injective.*

In particular, $\gamma(T)$ is a closed subspace of $L \cap \ker \partial_L$.

Proof

(i) $H(j)$ depends only on the homotopy class of γ.
(ii) Suppose that $H(\gamma)$ is not injective, and decompose $T = T(1) \oplus T(2)$ with $T(2) = \ker(T \to H(L))$. For each element x_i of a basis of $T(2)$, there is by definition an

element $y_i \in L$ such that $\gamma(x_i) = d(y_i)$. Denote $u_i = sx_i - y_i$. Then $du_i = 0$. Now let U be the vector space generated by the u_i. This induces a decomposition

$$(L \mathbin{\widehat{\amalg}} \overline{\mathbb{L}}_{sT}, \partial) \cong (L \mathbin{\widehat{\amalg}} \overline{\mathbb{L}}_{sT(1)}, \partial) \mathbin{\widehat{\amalg}} (\overline{\mathbb{L}}_U, 0).$$

It follows that

$$H(L \mathbin{\widehat{\amalg}} \overline{\mathbb{L}}_{sT}, \partial) \cong H(L \mathbin{\widehat{\amalg}} \overline{\mathbb{L}}_{sT(1)}, \partial) \mathbin{\widehat{\amalg}} \overline{\mathbb{L}}_U.$$

Since the image of $H(j)$ is contained in $H(L \mathbin{\widehat{\amalg}} \overline{\mathbb{L}}_{sT(1)}, \partial)$, the map $H(j)$ is not surjective.

$\square$

Next, denote by

$$\rho : \wedge W \to \mathbb{Q} \oplus S$$

a surjective Sullivan attaching map. Then set

$$(A, d) := \mathbb{Q} \oplus \ker \rho \subset \wedge W,$$

and let

$$\varphi : \wedge V \xrightarrow{\;\simeq\;} A$$

be a minimal Sullivan model. Finally we denote by $\lambda : \wedge V \to \wedge W$ the composite $\wedge V \xrightarrow{\simeq} A \to \wedge W$ and by $L_\lambda : L_W \longrightarrow L_V$ the induced morphism of homotopy Lie algebras.

Definition A Sullivan attaching map $\rho : \wedge W \to \mathbb{Q} \oplus S$ is *inert* if L_λ is surjective.

Now recall (Proposition 15.2) the bijection

$$[\overline{\mathbb{L}}_T, L] \longrightarrow [\wedge W, \mathbb{Q} \oplus S], \qquad \gamma \mapsto \rho$$

of homotopy classes of dgl attaching maps to homotopy classes of Sullivan attaching maps, in which γ is a profree dgl representative of ρ.

Proposition 16.2 *With the above notation, γ is inert $\Longleftrightarrow$ ρ is inert.*

Proof Let

$$\wedge V_E \to \wedge V_L$$

be the bihomogeneous morphism corresponding to the inclusion $L \to L\,\widehat{\amalg}\,\overline{\mathbb{L}}_{sT}$. By Proposition 15.3, this provides a homotopy commutative diagram

$$
\begin{array}{ccc}
\wedge V_E & \longrightarrow & \wedge V_L \\
\downarrow{\scriptstyle \simeq} & & \downarrow{\scriptstyle \simeq} \\
\wedge V & \longrightarrow & \wedge W
\end{array}
$$

in which $\wedge V$ is a minimal Sullivan algebra. This in turn identifies $H(V_E, d_0) \to H(V_L, d_0)$ with $V \to W$. Now Proposition 11.1(ii) identifies the morphism $H(L) \to H(L\,\widehat{\amalg}\,\overline{\mathbb{L}}_{sT})$ with the morphism $L_\lambda : L_W \to L_V$. $\qquad\square$

16.2   Relation with the Cofiber of an Attaching Map

Recall that the passage $X \mapsto X_{\mathbb{Q}}$ is carried out in two steps. The first one is the construction of the minimal Sullivan model of X, $(\wedge V, d)$, and the second is given by the geometric realization functor, $X_{\mathbb{Q}} = \langle \wedge V \rangle$.

For simply connected spaces, the passage preserves fibrations and cofibrations. In general it does not. For fibration, the Sullivan model of a fibration is not a fibration. However the geometric realization of a fibration (i.e., a Sullivan extension $\wedge V \to \wedge V \otimes \wedge W \to \wedge W$) is a fibration.

For cofibrations, the situation is different. The Sullivan model of a cofibration (example: $f : \vee S^{n_i} \to X$) is a cofibration. However the geometric realization does not preserve cofibrations, and if $\wedge V \to \wedge W$ is a Sullivan representative for $\iota : X \to X \cup_f (\vee_i D^{n_i+1})$, then the map $\langle \wedge W \rangle \to \langle \wedge V \rangle$ is not a model for the cofiber of $f_{\mathbb{Q}}$.

As an example we consider the injection of S^1 into $S^1 \vee S^1$ as its first component. The associated dgl model is the injection $\varphi : \overline{\mathbb{L}}_{(x)} \to \overline{\mathbb{L}}_{(x,y)}$. This is an inert attaching map, and the quotient is $\overline{\mathbb{L}}(y)$, whose geometric realization is $S^1_{\mathbb{Q}}$. On the other hand the homotopy cofiber Z of the map $S^1_{\mathbb{Q}} \to (S^1 \vee S^1)_{\mathbb{Q}}$ is more complicated. In fact since $H_2((S^1 \vee S^1)_{\mathbb{Q}}) \neq 0$ [54], $H_2(Z; \mathbb{Q}) \neq 0$, and so Z is very different from $S^1_{\mathbb{Q}}$.

16.3   Reduction to Quadratic Sullivan Algebras

Proposition 16.3

(i) *Let* $\rho : \wedge W \to \mathbb{Q} \oplus S$ *be a Sullivan attaching map. Regard* ρ *also as a morphism* $\rho_1 : (\wedge W, d_1) \to \mathbb{Q} \oplus S$, *where* $d_1 w$ *is the quadratic component of* dw, $w \in W$. *Then*

$$\rho \text{ is inert} \iff \rho_1 \text{ is inert}.$$

(ii) A morphism $\gamma : (\overline{\mathbb{L}}_T, 0) \to (L, \partial)$ of enriched dgl's is inert if and only if $H(\gamma)$: $(\overline{\mathbb{L}}_T, 0) \to (H(L), 0)$ is inert.

The proof of the proposition depends on a construction and on a key lemma. We establish these first and then provide the proof of the proposition. For the rest of the proof, (A, d) is a cdga with $A^0 = \mathbb{Q}$ and with a decomposition $A = \oplus_{p \geq 0} A^{(p)}$ in which each $A^{(p)}$ is a graded subspace. Assume further that

$$A^{(0)} = \mathbb{Q}, \, A^{(n)} \subset A^{\geq n}, \quad \text{and } A^{(p)} \cdot A^{(q)} \subset A^{(p+q)}$$

and that d decomposes as the sum of linear maps $d = \sum_{k \geq 1} d_k$ with $d_k : A^{(p)} \to A^{(p+k)}$.

In particular, (A, d_1) is a cdga equipped with a bigradation, one given by the usual degree and the second given by the decomposition $A = \oplus A^{(p)}$. It follows that (A, d_1) admits a bigraded minimal Sullivan model

$$\varphi_0 : (\wedge V, \delta_1) \to (A, d_1).$$

In particular, $V = \oplus_{p \geq 1} V^{(p)}$, $\wedge V = \oplus_{p \geq 0}(\wedge V)^{(p)}$ and

$$\delta_1 : V^{(p)} \to (\wedge V)^{(p+1)} \quad \text{and } \varphi_0 : (\wedge V)^{(p)} \to A^{(p)}.$$

We call this gradation the *new gradation*.

Lemma 16.2 *There is a differential d on $\wedge V$ and a morphism of cdga's $\varphi : (\wedge V, d) \to (A, d)$, such that*

$$d - \delta_1 : V^{(p)} \to (\wedge V)^{(>p+1)}$$

and

$$\varphi - \varphi_0 : V^{(p)} \to A^{(>p)}.$$

In particular, φ exhibits $(\wedge V, d)$ as a Sullivan model of (A, d).

Proof Starting with δ_1 and φ_0, we construct a sequence of derivations $\delta_2, \dots \delta_n, \dots$ in $\wedge V$ and a sequence of linear maps $\varphi_2, \varphi_3, \cdots : V \to A$ such that:

(i) δ_n increases the new gradation by n.
(ii) $\varphi_n(V^{(q)}) \subset A^{(n+q)}$.
(iii) Write $\varphi = \sum_{q \geq 0} \varphi_q$ and $d = \sum_{q \geq 1} \delta_q$. Then $d^2 = 0$, and $\varphi : (\wedge V, d) \to (A, d)$ is a morphism of cdga's.

Now we construct inductively $\delta_2, \ldots, \delta_n$ and $\varphi_1, \ldots, \varphi_{n-1}$ such that they satisfy the condition I_n:

$$\sum_{i+j=m} \delta_i \delta_k = 0, \quad 1 \le m \le n+1 \quad \text{and} \quad \sum_{j+k=m} \varphi_j \delta_k - d_k \varphi_j = 0, \quad 1 \le m \le n. \quad (I_n)$$

Since $(\wedge V, \delta_1)$ is a Sullivan model, we can write V as an increasing union of subspaces $V(n)$, with $V(0) = V \cap \ker \delta_1$ and $\delta_1 : V(p) \to \wedge V(p-1)$.

The first step is to define δ_2 on $V(0)$. Let $v \in V(0)^{(q)}$. Since $\delta_1 v = 0$, $d_1 \varphi_0 v = 0$, and therefore $d_1 d_2 \varphi_0(v) = 0$. Since φ_0 is a quasi-isomorphism, there are $a \in (\wedge V)^{(q+2)}$ and $b \in A^{(q+1)}$ such that $\delta_1 a = 0$ and $\varphi_0 a + d_1 b = d_2 \varphi_0(v)$. We define then $\delta_2 v = a$ and $\varphi_1 v = b$. This shows that the condition (I_2) is satisfied on $V(0)$.

Then, let $v \in V(p)$ for some $p > 0$. By induction $\delta_1 \delta_2 \delta_1(v) = -\delta_1^2 \delta_2(v) = 0$. On the other hand, $\varphi_0 \delta_2 \delta_1(v) = d_1(\varphi_1 \delta_1 - d_2 \varphi_0)(v)$. It follows that $\delta_2 \delta_1(v) = \delta_1(w)$, with $\varphi_0(w) = \varphi_1 \delta_1(v) - d_2 \varphi_0(v) + d_1 \gamma$ for some γ. We define $\delta_2(v) = -w$ and $\varphi_1(v) = -\gamma$. The condition (I_2) is thus satisfied.

Now, we supposed to have constructed by induction $\delta_2, \ldots, \delta_n$ and $\varphi_1, \ldots \varphi_{n-1}$ on V, satisfying the condition (I_n), and also δ_{n+1} and φ_n on $V(q)$ such that (I_{n+1}) is satisfied on $V(q)$. Then let $v \in V(q+1)$. By construction $\delta_{n+1} \delta_1(v)$ is already defined, and

$$\delta_1(\delta_{n+1}\delta_1 + \delta_n \delta_2 + \cdots + \delta_2 \delta_n) = -\sum_{j=1}^{n}\left(\sum_{r=2}^{n+3-j} \delta_r \delta_{n+3-j-r}\right)\delta_j$$

$$= -\sum_{r=2}^{n+1}\sum_{j=1}^{n+3-r} \delta_r\,(\delta_{n+3-j-r}\delta_j) = -\sum_{r=2}^{n+1}\left(\sum_{i+j=n+3-r} \delta_i \delta_j\right) = 0.$$

On the other hand, $\varphi_0(\delta_{n+1}\delta_1 + \delta_n \delta_2 + \cdots + \delta_2 \delta_n)$ is a d_1-boundary. Using induction we have

$$\varphi_0(\delta_{n+1}\delta_1 + \delta_n \delta_2 + \cdots + \delta_3 \delta_{n-1} + \delta_2 \delta_n)(v)$$

$$= \quad (-\varphi_1 \delta_n - \cdots - \varphi_n \delta_1)\delta_1(v) + (d_{n+1}\varphi_0 + d_n \varphi_1 + \cdots + d_1 \varphi_n)\delta_1(v)$$

$$+(-\varphi_1 \delta_{n-1} - \cdots - \varphi_{n-1}\delta_1)\delta_2(v) + (d_n \varphi_0 + \cdots + d_1 \varphi_{n-1})\delta_2(v) + \ldots$$

$$+(-\varphi_1 \delta_2 - \varphi_2 \delta_1)\delta_{n-1}(v) + (d_3 \varphi_0 + d_2 \varphi_1 + d_1 \varphi_0)\delta_{n-1}(v)$$

$$-\varphi_1 \delta_1 \delta_1 v + (d_2 \varphi_n + d_1 \varphi_1)\delta_n v$$

$$= -\varphi_1(\delta_n\delta_1 + \delta_{n-1}\delta_2 + \cdots + \delta_1\delta_n)(v) \cdots - \varphi_{n-1}(\delta_2\delta_1 + \delta_1\delta_2)(v) - \varphi_n\delta_1^2(v)$$

$$+d_{n+1}(\varphi_0\delta_1)(v) + d_n(\varphi_1\delta_1 + \varphi_0\delta_2)(v) + \cdots + d_1(\varphi_n\delta_1 + \varphi_{n-1}\delta_2 + \cdots + \varphi_1\delta_n)(v)$$

$$= d_{n+1}(d_1\varphi_0)(v) + d_n(d_2\varphi_0 + d_1\varphi_1)(v) + \cdots + d_1(\varphi_n\delta_1 + \varphi_{n-1}\delta_2 + \cdots + \varphi_1\delta_n)(v)$$

$$= (d_{n+1}d_1 + d_nd_2 + \cdots + d_2d_n)\varphi_0(v)$$

$$+(d_nd_1 + \cdots + d_2d_{n-1})\varphi_1(v) + \cdots + d_2d_1\varphi_{n-1}(v)$$

$$+d_1(\varphi_n\delta_1 + \varphi_{n-1}\delta_2 + \cdots + \varphi_1\delta_n)(v)$$

$$= -d_1(d_{n+1}\varphi_0 + d_n\varphi_1 + \cdots + d_2\varphi_{n-1})(v) + d_1(\varphi_n\delta_1 + \varphi_{n-1}\delta_2 + \cdots + \varphi_1\delta_n)(v)$$

Since φ_0 is a quasi-isomorphism, there are an element $w \in (\wedge V)^{(n+2)}$ and an element $z \in A^{(n+1)}$ such that $\delta_1 w = -(\delta_{n+1}\delta_1 + \cdots + \delta_2\delta_n)v$ and

$$\varphi_0(\delta_{n+1}\delta_1 + \cdots + \delta_2\delta_n)(v) + \varphi_0 w$$
$$= -d_{n+1}\varphi_0 - d_n\varphi_1 + \cdots - d_2\varphi_{n-1})(v) + (\varphi_n\delta_1 + \varphi_{n-1}\delta_2 + \cdots + \varphi_1\delta_n)(v) + d_1 z.$$

We define $\delta_{n+1}(v) = w$ and $\varphi_n(v) = z$. Then the above equation shows that the conditions (I_{n+1}) are satisfied for v.

Finally we define $d = \sum \delta_i$ and $\varphi = \sum \varphi_i$. Since $V^{(n)} \subset V^{\geq n}$ and $V = V^{\geq 1}$, the series are well defined. $\qquad\square$

Remark In Lemma 16.2, δ_1 is the part of the differential d that increases the gradation by one. In general, this is not the quadratic part d_1 of the differential.

Proof of Proposition 16.3

(i) Set $(A, d) = \mathbb{Q} \oplus \ker \rho = \mathbb{Q} \oplus (W \cap \ker \rho_{|W}) \oplus \wedge^{\geq 2} W$. Then set

$$A^{(1)} = \ker \rho_{|W} \quad \text{and} \quad A^{(p)} = \wedge^p W, \ p \geq 2.$$

It is immediate that the conditions of Lemma 16.2 are satisfied. This provides a Sullivan model $\varphi : (\wedge V, d) \overset{\simeq}{\longrightarrow} (A, d)$, with $\varphi = \varphi_0 + \varphi_1 + \ldots, d = \delta_1 + \delta_2 + \ldots$, and such that

$$\varphi_0 : (\wedge V, \delta_1) \overset{\simeq}{\longrightarrow} (A, d_1)$$

is also a Sullivan model and $\varphi - \varphi_0 : V \to A^{\geq 2)} = \wedge^{\geq 2} W$.

By definition, ρ is inert if and only if φ induces an injection $V \to A/\wedge^{\geq 2} W$. This is equivalent to φ_0 inducing an injection $V \to A/\wedge^{\geq 2} W$. But $(A, d_1) = \ker \rho_1$ and so this is equivalent to ρ_1 being inert.

(ii) Denote by $(\wedge W, d)$ the minimal Sullivan algebra associated with (L, ∂). Let $(\wedge V_{H(L)}, d_1)$ be the quadratic Sullivan algebra corresponding to the enriched Lie algebra $H(L)$. Then combining Lemma 11.1(ii) and Proposition 11.1 yields isomorphisms

$$(\wedge V_{H(L)}, d_1) \xrightarrow{\cong} (\wedge H(V_L, d_0), \overline{d_1}) \xrightarrow{\cong} (\wedge W, d_1).$$

It follows that $H(\gamma)$ is inert if and only if ρ_1 is inert. $\qquad\square$

16.4 A Sullivan Characterization of Inert Maps

To fix the notation, we denote by $\rho : \wedge W \to \mathbb{Q} \oplus S$ a surjective cdga morphism from a minimal Sullivan algebra. Division by $\mathbb{Q}$ converts ρ to a surjective map $\overline{\rho} : W \to S$. We extend $\overline{\rho}$ into a surjective map of complexes $\widehat{\rho} : \wedge W \to S$ with $\overline{\rho}(\mathbb{Q} \oplus \wedge^{\geq 2} W) = 0$.

Let $(A, d) = \mathbb{Q} \oplus \ker \overline{\rho} \subset \wedge W$, and let $\varphi : \wedge V \xrightarrow{\simeq} A$ be a minimal Sullivan model. Then denote by $\lambda : \wedge V \to \wedge W$ the composite $\wedge V \xrightarrow{\simeq} A \to \wedge W$ and by

$$L_\lambda : L_W \longrightarrow L_V$$

the induced morphism of homotopy Lie algebras.

This yields the row-exact commutative diagram of complexes

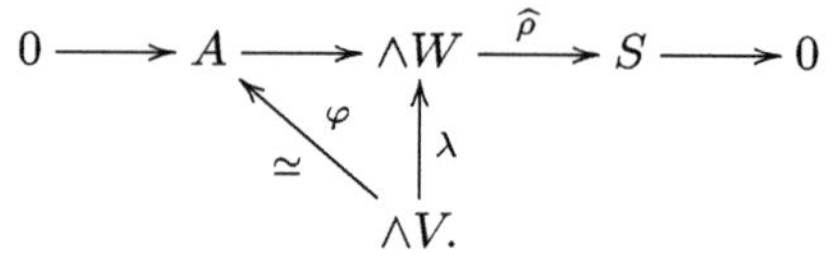

$$
\begin{array}{ccccccccc}
0 & \longrightarrow & A & \longrightarrow & \wedge W & \xrightarrow{\widehat{\rho}} & S & \longrightarrow & 0 \\
& & & \nwarrow{\scriptstyle \varphi} & \uparrow{\scriptstyle \lambda} & & & & \\
& & & \simeq & \wedge V. & & & &
\end{array}
$$

Since $\widehat{\rho}$ is surjective, the exact sequence

$$0 \to H^1(A) \to H^1(\wedge W) \to S^1 \to 0$$

shows that $H^1(\lambda)$ is injective. Therefore λ factors as [36, Theorem 3.1]

$$\lambda : \wedge V \xrightarrow{\ \eta\ } \wedge V \otimes \wedge Z \xrightarrow[\simeq]{\ \xi\ } \wedge W \ , \qquad \eta(v) = v \otimes 1,$$

in which η is a minimal Λ-extension. Denote the quotient minimal Sullivan algebra by $(\wedge Z, \overline{d})$. Finally, let $\wedge V \otimes \wedge U$ be the acyclic closure of $\wedge V$.

Proposition 16.4 *With the above notation,*

$$H(\wedge Z) \cong \mathbb{Q} \oplus (S \otimes \wedge U).$$

Proof Applying $- \otimes_{\wedge V} \wedge V \otimes \wedge U$ to the short exact sequence of complexes

$$0 \to A \to \wedge W \xrightarrow{\ \widehat{\rho}\ } S \to 0$$

and to the quasi-isomorphism ξ yields the isomorphisms

$$
\begin{array}{ccccccccc}
0 & \longrightarrow & A \otimes_{\wedge V}(\wedge V \otimes \wedge U) & \longrightarrow & \wedge W \otimes_{\wedge V}(\wedge V \otimes \wedge U) & \longrightarrow & S \otimes_{\wedge V}(\wedge V \otimes \wedge U) & \longrightarrow & 0 \\
& & \downarrow{\scriptstyle\simeq} & & \downarrow{\scriptstyle\simeq} & & \downarrow{\scriptstyle\simeq} & & \\
& & \mathbb{Q} & & \wedge Z & & (S \otimes \wedge U, 0). & &
\end{array}
$$

Therefore,

$$H(\wedge Z) \cong \mathbb{Q} \oplus (S \otimes \wedge U).$$

$\square$

Remark The isomorphism of Proposition 16.4 is a generalization of the following result of [31]: Let F be the homotopy fiber of a map $X \to Y := X \cup_g (\vee_\alpha D_\alpha^{n+1})$ with X simply connected and $n \geq 1$. The morphism $g : S = \vee_\alpha S_\alpha^n \to X$ then lifts to a map $S \to F$. Denote by $u_\alpha \in H_n(F; \mathbb{Q})$ the image of the fundamental class of S_α^n. The holonomy action of $H_*(\Omega Y; \mathbb{Q})$ on $H_*(F; \mathbb{Q})$ then yields an isomorphism

$$H_{\geq 1}(F; \mathbb{Q}) \cong (\oplus_\alpha \mathbb{Q} u_\alpha) \otimes H_*(\Omega Y; \mathbb{Q}) \cong H^{\geq 1}(S) \otimes H_*(\Omega Y;).$$

Theorem 16.1 *With the hypotheses and the notation above, the following conditions are equivalent:*

(i) $\rho : \wedge W \to \mathbb{Q} \oplus S$ is inert.
(ii) $\wedge V \otimes \wedge Z$ is a minimal Sullivan algebra.
(iii) The Lie algebra $L_{\wedge Z}$ is profree.

Proof (i) $\Longleftrightarrow$ (ii). By definition, ρ is inert if and only if $L_W \to L_V$ is surjective, i.e., if $V \to \wedge^{\geq 1} W / \wedge^{\geq 2} W$ is injective. But this map is an inclusion if and only if $\lambda : \wedge V \to \wedge V \otimes \wedge Z$ is a minimal Λ-extension.

(ii) $\Longrightarrow$ (iii) Identify $\wedge V \otimes \wedge Z = \wedge W$ via the isomorphism ξ, and denote by $\wedge V \otimes \wedge U$ the acyclic closure of $\wedge V$. Then, applying $- \otimes_{\wedge V} \wedge V \otimes \wedge U$ to the quasi-isomorphism ξ yields a quasi-isomorphism

$$\tau : (\wedge W \otimes \wedge U, d) \xrightarrow{\sim} (\wedge Z, \overline{d}).$$

We show that τ has a right inverse

$$\sigma : (\wedge Z, \overline{d}) \to (\wedge W \otimes \wedge U, d)$$

in which $\sigma : Z \to \wedge^{\geq 1} W \otimes \wedge U$.

In fact, write $Z = \cup_k Z_k$ with $Z_{k+1} = Z \cap d^{-1}(\wedge V \otimes \wedge Z_k)$. Then assume by induction that σ has been constructed in some Z_k with $\sigma(Z_k) \subset \wedge^{\geq 1} W \otimes \wedge U$. This implies that for $z \in Z_{k+1}$

$$\sigma(\overline{d}z) \in \sigma(\wedge^{\geq 2} Z_k) \subset \wedge^{\geq 2} W \otimes \wedge U.$$

In particular, $\sigma(\overline{d}z)$ is a cycle in $A^{\geq 1} \otimes \wedge U$ which is quasi-isomorphic to $\wedge^{\geq 1} V \otimes \wedge U$. Since $H^{\geq 1}(\wedge V \otimes \wedge U) = 0$, $\sigma(\overline{d}z) = d\Phi$ for some $\Phi \in \wedge W \otimes \wedge U$.

Next write $\Phi = \Phi_1 + \Phi_2$ with $\Phi_1 \in \wedge^{\geq 1} W \otimes \wedge U$ and $\Phi_2 \in \wedge U$. Since $d : U \to \wedge^{\geq 1} V \otimes \wedge U$ and $\wedge W$ is minimal, $d\Phi_1 \in \wedge^{\geq 2} W \otimes \wedge U$. Since $d\Phi = \sigma(\overline{d}z) \in \wedge^{\geq 2} W \otimes \wedge U$, we have $d\Phi_2 \in \wedge^{\geq 2} W \otimes \wedge U$. But as with any acyclic closure, if $\Phi_2 \neq 0$, then $d\Phi_2$ has a nonzero component in $V \otimes \wedge U$. Thus $\Phi_2 = 0$ and $\sigma(\overline{d}z) = d\Phi_1$. Extend σ to Z_{k+1} by setting $\sigma(z) = \Phi_1$ for a basis element z of a direct summand of Z_k in Z_{k+1}.

Finally, apply $- \otimes_{\wedge V} \wedge V \otimes \wedge U$ to the exact sequence of complexes $0 \to A \to \wedge W \to S \to 0$. This yields the commutative diagram

$$
\begin{array}{ccccccccc}
0 & \longrightarrow & \mathbb{Q} & \longrightarrow & \wedge Z & \longrightarrow & \wedge^{\geq 1} Z & \longrightarrow & 0 \\
& & {\scriptstyle \simeq}\downarrow{\scriptstyle \alpha} & & {\scriptstyle \sigma}\downarrow{\scriptstyle \simeq} & & {\scriptstyle \overline{\sigma}}\downarrow{\scriptstyle \simeq} & & \\
0 & \longrightarrow & A \otimes \wedge U & \longrightarrow & \wedge W \otimes \wedge U & \xrightarrow{\widehat{\rho}\otimes id} & S \otimes \wedge U & \longrightarrow & 0,
\end{array}
$$

in which α is a quasi-isomorphism because $\wedge V \xrightarrow{\sim} A$. It follows that $\overline{\sigma}$ is a quasi-isomorphism in which $\overline{\sigma}(\wedge^{\geq 2} Z) = 0$. In particular, the injection $Z \cap \ker \overline{d} \to \wedge^{\geq 1} Z$ is a quasi-isomorphism, and $\mathbb{Q} \oplus (Z \cap \ker \overline{d}) \xrightarrow{\sim} \wedge Z$. It follows from Theorem 6.1 that $L_{\wedge Z}$ is profree.

(iii) $\implies$ (ii) Here, even if $\wedge V \otimes \wedge Z$ is not a minimal Sullivan algebra, we have a row-exact commutative diagram of the form

$$
\begin{array}{ccccccccc}
0 & \longrightarrow & B & \longrightarrow & \wedge V \otimes \wedge Z & \xrightarrow{\widehat{\rho}\circ\xi} & S & \longrightarrow & 0 \\
& & \downarrow{\scriptstyle \psi} & & {\scriptstyle \simeq}\downarrow{\scriptstyle \xi} & & \parallel{\scriptstyle =} & & \\
0 & \longrightarrow & A & \longrightarrow & \wedge W & \xrightarrow{\widehat{\rho}} & S & \longrightarrow & 0,
\end{array}
$$

where $B = \ker(\widehat{\rho} \circ \xi)$. Since $\widehat{\rho} \circ \overline{\xi}(V) = 0$, $\wedge V \otimes 1 \subset B$. Thus, since ψ is surjective, $B \to A$ is a quasi-isomorphism, and therefore so is the inclusion $\wedge V \to B$.

Now spatial realization provides a fibration [33, Proposition 17.9]

$$\langle \wedge Z \rangle \longrightarrow \langle \wedge V \otimes \wedge Z \rangle \longrightarrow \langle \wedge V \rangle,$$

with a connecting homomorphism $\partial : \pi_* \langle \wedge V \rangle \to \pi_{*-1} \langle \wedge Z \rangle$. By Proposition 9.9, $\partial \alpha$ is in the center of the profree Lie algebra L_Z. This implies that $L_Z = \overline{\mathbb{L}}_x$ and $\dim H^{\geq 1}(\wedge Z) = 1$.

On the other hand, by Proposition 16.2, $H^{\geq 1}(\wedge Z) \cong S \otimes \wedge U$. Thus if $\dim H^{\geq 1}(\wedge Z) = 1$, then $U = 0$, $\wedge V \otimes \wedge Z = \wedge Z$, and $\wedge V \otimes \wedge Z$ is minimal. $\square$

Corollary *The following conditions on the homotopy Lie algebra, L_W, of a minimal Sullivan algebra, $\wedge W$, are equivalent:*

(i) L_W is profree: $L_W = \overline{\mathbb{L}}_T$, where T is any closed direct summand of $L_W^{(2)}$.
(ii) Each closed direct summand, T, of $L_W^{(2)}$ is inert.
(iii) Some closed direct summand, T, of $L_W^{(2)}$ is inert.

Proof (i) $\Rightarrow$ (ii) By Proposition 8.6 we may assume that $(\wedge W, d)$ is quadratic and that $W \cap \ker d \xrightarrow{\cong} H^{\geq 1}(\wedge W)$. Moreover, the inclusion of T in L_W is dual to a surjection $\rho : W \to W \cap \ker d$. Since $\wedge W \xrightarrow{\cong} \mathbb{Q} \oplus (W \cap \ker d)$, $H(A) = \mathbb{Q}$, $V = 0$, and T is inert.

(ii) $\Rightarrow$ (iii) is immediate.

(iii) $\Rightarrow$ (i). Let S be the dual of sT and $\rho : \wedge W \to S \oplus \mathbb{Q}$ dual to the injection $T \to L_W$. As usual we denote by $\wedge V$ a minimal Sullivan model of $\mathbb{Q} \oplus \ker \rho$ and by $\varphi : \wedge V \to \wedge W$ the induced map. Since ρ is inert, φ is injective and induces an injection at the level of indecomposable elements $V \to W$. But $\rho : W \to S$ is an isomorphism, so that $V = 0$ and $\wedge W \xrightarrow{\cong} \mathbb{Q} \oplus S$. This implies the result by Theorem 6.1. $\square$

16.5 Holonomy Representations and Inert Maps

In this subsection we fix an inert morphism $\rho : \wedge W \to \mathbb{Q} \oplus S$, and we consider the associated diagram

$$
\begin{array}{ccccc}
\wedge V & \longrightarrow & \wedge V \otimes \wedge Z & \longrightarrow & \wedge Z \\
\downarrow{\varphi} & & \cong \downarrow{\xi} & & \downarrow{\rho_Z} \\
A = \mathbb{Q} \oplus \ker \rho & \longrightarrow & \wedge W & \xrightarrow{\rho} & \mathbb{Q} \oplus S.
\end{array}
$$

By Theorem 16.1, $L_Z = \overline{\mathbb{L}}_R$ is profree, and we denote by $\gamma : T \to L_Z$ the inclusion dual to the surjection $\rho_Z : \wedge Z \to \mathbb{Q} \oplus S$.

Proposition 16.5 *When ρ is inert, there is a commutative diagram*

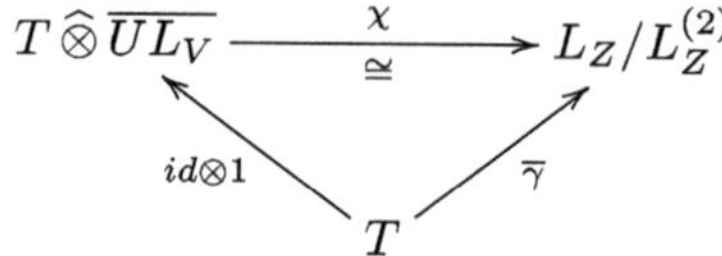

in which χ is an isomorphism of $\overline{UL_V}$-modules, and:

- The representation in $L_Z/L_Z^{(2)}$ is induced by the right adjoint representation of $\overline{UL_V}$ in L_Z.
- The representation in $T \widehat{\otimes} \overline{UL_V}$ is multiplication from the right.

Proof By Proposition 16.3 we can suppose that the differential d in $\wedge W$ is quadratic.

The proof is in several steps. Throughout, $\varepsilon_V \otimes \varepsilon_U$ denotes the augmentation in $\wedge V \otimes \wedge U$, with $\varepsilon_U(U) = 0$.

Step 1. *The morphism*

$$\varepsilon_V \otimes \rho_Z : (\wedge V \otimes (Z \cap \ker d_Z), d) \xrightarrow{\simeq} (S, 0)$$

is a quasi-isomorphism.

Since $\wedge W$ is a quadratic Sullivan algebra, it follows that

$$d : Z \cap \ker d_Z \to \wedge^2 V \oplus (V \otimes (Z \cap \ker d_Z)).$$

Thus division by the subcomplex $\wedge V \otimes 1$ defines a quadratic $\wedge V$-module $(\wedge V \otimes (Z \cap \ker d_Z), d)$. In particular,

$$\varepsilon_V \otimes \rho_Z : \wedge V \otimes (Z \cap \ker d_Z) \to S$$

is a morphism of chain complexes.

On the other hand, the row-exact commutative diagram

$$
\begin{array}{ccccccccc}
0 & \longrightarrow & (\wedge V, d) & \longrightarrow & (\wedge V \otimes \wedge Z, d) & \longrightarrow & (\wedge V \otimes \wedge^{\geq 1} Z, \bar{d}) & \longrightarrow & 0 \\
 & & \downarrow{\simeq} & & \downarrow{\cong} & & \downarrow{\simeq} & & \\
0 & \longrightarrow & A & \longrightarrow & \wedge W & \longrightarrow & S & \longrightarrow & 0
\end{array}
$$

shows that the induced map $\rho : \wedge V \otimes \wedge^{\geq 1} Z \to S$ is a quasi-isomorphism. But since ρ is inert, L_Z is profree (Theorem 16.1). Therefore (Theorem 6.1) $Z \cap \ker d_Z \to (\wedge^{\geq 1} Z, d_Z)$ is a quasi-isomorphism. The sequence of quasi-isomorphisms

$$\varepsilon_V \otimes \rho_Z : \wedge V \otimes (Z \cap \ker d_Z) \xrightarrow{\simeq} \wedge V \otimes \wedge^{\geq 1} Z \xrightarrow[\rho]{\simeq} S \, ,$$

completes the proof of Step 1.

Step 2. *There is a quasi-isomorphism of* $\wedge V$-*quadratic modules*

$$\psi : \wedge V \otimes (Z \cap \ker d_Z) \xrightarrow{\simeq} (\wedge V \otimes \wedge U) \otimes S.$$

Here $\wedge V \otimes \wedge U$ is the acyclic closure of $\wedge V$, and (Sect. 5.4) is a quadratic $\wedge V$-module. Thus $(\wedge V \otimes \wedge U) \otimes S$ is also a quadratic $\wedge V$-module, and

$$\varepsilon_V \otimes \varepsilon_U \otimes id : \wedge V \otimes \wedge U \otimes S \xrightarrow{=} S$$

is a quasi-isomorphism. Since the quadratic $\wedge V$-modules are semifree, the quasi-isomorphism $\varepsilon_V \otimes \rho_Z$ of Step 1 lifts through $\varepsilon_V \otimes \varepsilon_U \otimes id$ to yield a commutative diagram of $\wedge V$-modules

$$
\begin{array}{ccc}
\wedge V \otimes (Z \cap \ker d_Z) & \xrightarrow[\simeq]{\psi} & \wedge V \otimes \wedge U \otimes S \\
& {\scriptstyle \varepsilon_V \otimes \rho_Z} \searrow \;\; \simeq \qquad \simeq \;\; \swarrow {\scriptstyle \varepsilon_V \otimes \varepsilon_U \otimes id} & \\
& S &
\end{array}
\tag{16.1}
$$

Indeed assign to any quadratic $\wedge V$-module a secondary gradation with $\wedge^k V \otimes -$ having secondary degree k. Then, if S is assigned secondary degree one, $\varepsilon_V \otimes \rho_Z$ and $\varepsilon_V \otimes \varepsilon_U \otimes id$ preserve the secondary degree, while the differentials increase it by 1. It follows that ψ can be constructed to be a morphism of $\wedge V$-modules preserving secondary degree. This implies that ψ is a quasi-isomorphism of quadratic $\wedge V$-modules.

Step 3. *Completion of the proof of Proposition 16.5*

It follows from Lemma 5.1(ii) that the induced map $\overline{\psi}$ is an isomorphism and that the diagram (16.1) restricts to a commutative diagram

$$
\begin{array}{ccc}
Z \cap \ker d_Z & \xrightarrow[\cong]{\overline{\psi}} & \wedge U \otimes S \\
& {\scriptstyle \rho_Z} \searrow \qquad \swarrow {\scriptstyle \varepsilon_U \otimes id} & \\
& S. &
\end{array}
$$

Now the tensor product $\psi \otimes_{\wedge V} (\wedge V \otimes \wedge U)$ induces a morphism ψ' and a commutative diagram

$$
\begin{array}{ccc}
Z \cap \ker d_Z & \xrightarrow[\cong]{\overline{\psi}} & \wedge U \otimes S \\
{\scriptstyle \simeq} \uparrow & & \uparrow {\scriptstyle \simeq} \\
\wedge V \otimes (Z \cap \ker d_Z) \otimes \wedge U & \xrightarrow[\cong]{\psi'} & (\wedge V \otimes \wedge U \otimes S) \otimes_{\wedge V} (\wedge V \otimes \wedge U).
\end{array}
$$

Denote by $\sigma : \wedge U \to \wedge V \otimes \wedge U \otimes \wedge U$ a quasi-isomorphism inverse to the projection $\varepsilon_V \otimes id \otimes id$. This gives the commutative diagram

$$
\begin{array}{ccc}
Z \cap \ker d_Z & \xrightarrow{\ \overline{\psi}\ } & \wedge U \otimes S \\
\downarrow{\scriptstyle \sigma'} & & \downarrow{\scriptstyle \sigma \otimes id} \\
\wedge V \otimes (Z \cap \ker d_Z)) \otimes \wedge U & \xrightarrow{\ \psi'\ } & \wedge V \otimes \wedge U \otimes \wedge U \otimes S \\
\downarrow{\scriptstyle \varepsilon_V} & & \downarrow{\scriptstyle \varepsilon_V} \\
(Z \cap \ker d_Z) \otimes \wedge U & \xrightarrow{\ \overline{\psi} \otimes 1\ } & \wedge U \otimes \wedge U \otimes S,
\end{array}
$$

where $\sigma' = (\psi')^{-1}(\sigma \otimes 1)\overline{\psi}$.

The dual of the right vertical composition is the morphism

$$
\overline{UL_V} \,\widehat{\otimes}\, \overline{UL_V} \,\widehat{\otimes}\, T \xrightarrow{\ \mu \otimes 1\ } \overline{UL_V} \,\widehat{\otimes}\, T,
$$

where μ is the multiplication in $\overline{UL_V}$.

The composition $\varepsilon_V \circ \sigma'$ gives to $L_Z/L_Z^{(2)}$ the structure of a $\overline{UL_V}$-module. Dualizing this yields the commutative diagram of $\overline{UL_V}$-modules

$$
\begin{array}{ccc}
L_Z/L_Z^{(2)} = s^{-1}(Z \cap \ker d_Z)^\vee & \xleftarrow{\ \cong\ } & T \widehat{\otimes} \overline{UL_V} \\
& {\scriptstyle \overline{\gamma}} \nwarrow \quad \nearrow {\scriptstyle id \otimes 1} & \\
& T &
\end{array}
$$

in which $\overline{UL_V}$ acts by right multiplication in $T \,\widehat{\otimes}\, \overline{UL_V}$. $\qquad\qquad \square$

16.6   A Holonomy Characterization for Inert Maps

Theorem 16.2 *Let $\gamma : \overline{\mathbb{L}}_T \to L$ be a morphism of enriched dgl's. Then the following conditions on γ are equivalent:*

(i) γ is inert.

(ii) The closed ideal I in $H(L)$ generated by $\gamma(T)$ is profree, and the right adjoint representation of L/I in $I/I^{(2)}$ induces an isomorphism of $\overline{UL/I}$-modules

$$
T \,\widehat{\otimes}\, \overline{UL/I} \xrightarrow{\ \cong\ } I/I^{(2)}.
$$

Moreover, when γ is inert, the ideal I generated by $\gamma(T)$ coincides with the kernel of the projection $H(L) \to H(L \,\widehat{\amalg}\, \overline{\mathbb{L}}_{sT})$.

Proof (i) $\Rightarrow$ (ii). Since γ is inert, the map $H(L) \to H(L \,\widehat{\amalg}\, \overline{\mathbb{L}}_{sT})$ is surjective, and by Theorem 16.1, its kernel L_Z is a free Lie algebra, $L_Z = \overline{\mathbb{L}}_R$. We denote by $\overline{\gamma}$ the factorization of γ through L_Z. This gives the commutative diagram in which the horizontal line is short exact

$$0 \longrightarrow L_Z = \overline{\mathbb{L}}_R \longrightarrow H(L) \longrightarrow H(L \,\widehat{\amalg}\, \overline{\mathbb{L}}_{sT}) \longrightarrow 0$$

Proposition 16.5 shows that $\overline{\gamma}(T)$ generates $L_Z/L_Z^{(2)}$. Therefore $I = L_Z$ and $I/I^{(2)} = L_Z/L_Z^{(2)}$. Now (ii) follows from Proposition 16.5.

(ii) $\Rightarrow$ (i). Since $I \subset L$ is a closed ideal, it decomposes $\wedge W$ as a minimal Λ-extension

$$\wedge W = \wedge V_I \otimes \wedge Z_I$$

in which $V_I = \{w \in W \mid\, < w, sI > = 0\}$. In particular, $I = L_{Z_I}$, and, since

$$< \rho V_I, sT > = < V_I, s\gamma(T) > = 0,$$

it follows that $\wedge V_I \subset A$ and $W \to S$ factorizes through a map $\rho_{Z_I} : Z_I \to S$.

Since $\wedge W$ is a quadratic Sullivan algebra, it follows that division by $\wedge V_I \otimes 1$ defines a quadratic $\wedge V_I$-module $(\wedge V_I \otimes (Z_I \cap \ker d_{Z_I}), d)$. In particular,

$$\varepsilon_{V_I} \otimes \rho_{Z_I} : \wedge V_I \otimes (Z_I \cap \ker d_{Z_I}) \to S$$

is a morphism of chain complexes.

Now let $\wedge V_I \otimes \wedge U_I$ be the acyclic closure of $\wedge V_I$ and is a quadratic $\wedge V_I$-module. Thus $\wedge V_I \otimes \wedge U_I \otimes S$ is also a quadratic $\wedge V_I$-module, and

$$\varepsilon_{V_I} \otimes \varepsilon_{U_I} \otimes id : \wedge V_I \otimes \wedge U_I \otimes S \xrightarrow{\;=\;} S$$

is a quasi-isomorphism. Since the quadratic $\wedge V_I$-modules are semifree, the quasi-isomorphism $\varepsilon_{V_I} \otimes \rho_{Z_I}$ lifts through $\varepsilon_{V_I} \otimes \varepsilon_{U_I} \otimes id$ to yield a commutative diagram of $\wedge V_I$-modules

$$\wedge V_I \otimes (Z_I \cap \ker d_{Z_I}) \xrightarrow{\;\;\psi\;\;} \wedge V_I \otimes \wedge U_I \otimes S$$

$$\varepsilon_{V_I} \otimes \rho_{Z_I} \searrow \quad \swarrow \varepsilon_{V_I} \otimes \varepsilon_{U_I} \otimes id$$

$$S$$

We deduce the commutative diagram

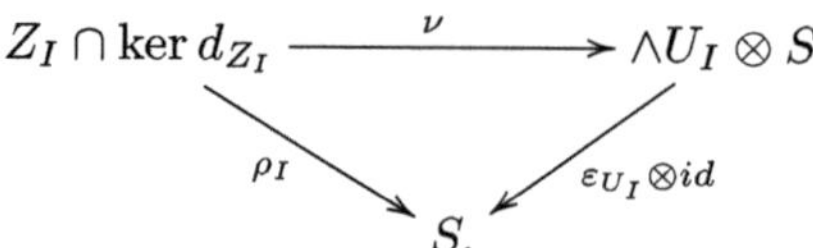

Then ν dualizes in a morphism of $\overline{UL_{V_I}}$-modules, and the diagram dualizes to

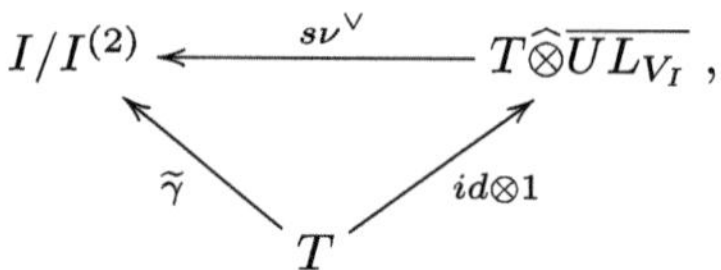

in which $\widetilde{\gamma}$ is the composite $T \xrightarrow{\gamma} I \to I/I^{(2)}$. Now $s\nu^\vee$ is, by hypothesis, an isomorphism of the $\overline{UL_{V_I}}$-modules. Therefore ν and ψ are isomorphisms, and it follows (because I is profree) that $\wedge V_I \otimes \wedge^{\geq 1} Z_I \xrightarrow{\cong} S$. Now from the injection $\wedge V_I \subset A$, we have the following commutative diagram:

$$
\begin{array}{ccccccccc}
0 & \longrightarrow & (\wedge V_I, d) & \longrightarrow & (\wedge V_I \otimes \wedge Z_I, d) & \longrightarrow & (\wedge V_I \otimes \wedge^{\geq 1} Z_I, \bar{d}) & \longrightarrow & 0 \\
 & & \downarrow & & \downarrow{\scriptstyle\cong} & & \downarrow{\scriptstyle\simeq} & & \\
0 & \longrightarrow & A & \longrightarrow & \wedge W & \longrightarrow & S & \longrightarrow & 0
\end{array}
$$

It follows that $\wedge V_I \xrightarrow{\cong} A$.

But by definition, $\varphi : \wedge V \xrightarrow{\cong} A$ is a minimal Sullivan model. Therefore we can choose φ so that it factors as

$$\wedge V \xrightarrow{\cong} \wedge V_I \to A.$$

In particular, φ extends to an isomorphism $\wedge V \otimes \wedge Z_I \xrightarrow{\cong} \wedge W$, and, by Theorem 16.1, this implies that ρ is inert. $\qquad\square$

Corollary *If $\gamma : \mathbb{Q}x \to L$ is an inert attaching map and if $y \in L$ is not a scalar multiple of x or of $[x, x]$, then $[x, y] \neq 0$.*

Proof Let I be the closed ideal generated by x. Then (Theorem 16.1) I is profree. If $y \in I$ and if $y \notin \mathbb{Q}x + \mathbb{Q}[x, x]$, then the sub-Lie algebra generated by x and y is free, and $[x, y] \neq 0$.

On the other hand when $y \notin I$, let $\bar{y} = [y] \in L/I$. Since $I/I^{(2)}$ is isomorphic to $x \otimes \overline{U(L_W/I)}$ via the adjoint representation, $x \cdot \bar{y} = \overline{[x, y]}$ is nonzero. $\qquad\square$

Remark Let $f : X \to Y$ be a continuous map from a wedge of spheres X. When f is inert, the homotopy of the cofiber X depends only on the map $\pi_*(X) \to \pi_*(Y)$. This is not the case in general.

As an example denote by $f : S^2 \to \mathbb{C}P^3$ the injection of the 2-skeleton in $\mathbb{C}P^3$. The homotopy cofiber is $Z = \mathbb{S}^4 \vee S^6$ with homotopy Lie algebra $L_Z = \mathbb{L}(x_3, x_5)$. Now let $g : S^2 \to S^3 \times \mathbb{C}P^\infty$ denote also the injection of the 2-skeleton. The maps $\pi_*(f) \otimes \mathbb{Q}$ and $\pi_*(g) \otimes \mathbb{Q}$ coincide. The homotopy cofiber of g is a formal space Z' whose cohomology is

$$\wedge(x, y, z, t)/(x^3 - y^2, zt, tx - yz, ty - x^2 b),\ deg\, x = 4,\ deg\, y = 6,\ deg\, z = 7,\ deg\, t = 9.$$

The homotopy Lie algebras L_Z and $L_{Z'}$ are clearly different.

Examples of Inert Subspaces

(1) The fact for a closed ideal I to be profree is not enough for I to be inert. Consider, for instance, the space $X = S^1 \times (S_a^3 \vee S_b^3)$. The sphere S_a^3 generates a profree Lie subalgebra in $\pi_*(X) \otimes \mathbb{Q}$ but is not inert. Indeed,

$$X \cup_{S_a^3} e^4 \cong (S^1 \times S_b^3) \vee S^4.$$

(2) Let X and Y be wedges of spheres with profree models $(\overline{\mathbb{L}}_{(x_i, i \in I)}, 0)$ and $(\overline{\mathbb{L}}_{(y_j, j \in J)}, 0)$. We denote by T the vector space generated by the Lie brackets $[x_i, y_j]$. Then it follows from Proposition 8.7 that T is an inert subspace in $L_{X \vee Y}$ with quotient $L_{X \times Y}$.

In particular, the kernel of $L_{X \vee Y} \to L_{X \times Y}$ is the free Lie algebra L_R with

$$R = T \,\widehat{\otimes}\, \overline{UL_X} \,\widehat{\otimes}\, \overline{UL_Y},$$

the action being given by iterative adjonctions.

General Comment

Let $\gamma : (\overline{\mathbb{L}}_T, 0) \to (L, \partial)$ be a morphism of enriched dgl's, and let $\varphi : (L, \partial) \to (L \,\widehat{\amalg}\, \overline{\mathbb{L}}_{sT}, \delta)$ be the associated attachment map, with $\partial(st) = \gamma(t)$. Denote by I the ideal in $H(L)$ generated by $\gamma(T)$, and let $E = H(L)/I$. Finally denote by J the kernel of the map $H(L) \to H(L \,\widehat{\amalg}\, \overline{\mathbb{L}}_{sT})$.

When γ is inert, $I = J$, and I is a profree Lie algebra. In the general case, this can be very different. For instance, suppose that γ is the profree dgl model of the inclusion $S^2 \to \mathbb{C}P^2$. Then $\pi_*\Omega\mathbb{C}P^2 \otimes \mathbb{Q}$ is the abelian Lie algebra on two generators, a in degree 1 and b in degree 4. Here $I = \mathbb{Q}a$ and $J = \mathbb{Q}a \oplus \mathbb{Q}b$. Moreover I is not profree.

When (L, ∂) has the form $(\overline{\mathbb{L}}_V, 0)$, then $I = J$, and I is a profree Lie algebra.

When γ is inert, $I = \overline{\mathbb{L}}_R$, and R is the profree $\overline{U E}$-module generated by $\gamma(T)$. In general I is always generated as a Lie algebra by the graded vector space $R = \overline{U E} \,\widehat{\otimes}\, T$, but the structure of I can be more complicated. In particular the intersection $R \cap \mathbb{L}^2(R)$ can be non-empty. Consider, for instance, the situation where $(L, \partial) = (\overline{\mathbb{L}}(x, y, z, t), 0)$

and $\gamma(T)$ is generated by the six brackets of length 2 in the variables x, y, z, t. In this case $[[x, y], [z, t]]$ belongs to the $\overline{UE}$-module generated by the other basic brackets.

Open Questions

1. Suppose that α is an inert element in an enriched Lie algebra L and that β belongs to the closed ideal generated by α, is β also an inert element?
2. Let $X = Y \vee Z$ be the wedge of two connected spaces, and let F be the homotopy fiber of the injection $(Y \vee Z)_{\mathbb{Q}} \to (Y \times Z)_{\mathbb{Q}}$. Then $L_{Y \vee Z} \cong L_Y \,\widehat{\amalg}\, L_Z$ (Proposition 10.2) and $L_{Y \times Z} \xrightarrow{\simeq} L_Y \oplus L_Z$. Moreover L_F is a profree Lie algebra (Proposition 10.4). This gives the short exact sequence

$$0 \to L_F \to L_Y \,\widehat{\amalg}\, L_Z \to L_Y \oplus L_Z \to 0.$$

We wonder if a nonzero element in L_F is an inert element in $L_Y \,\widehat{\amalg}\, L_Z$.

16.7   Freely Generated Ideals

The prototype of an inert map is an injection of the form

$$\gamma : (\overline{\mathbb{L}}_T, 0) \to (\overline{\mathbb{L}}_T, 0) \,\widehat{\amalg}\, (L', \partial) = (L, \partial).$$

In this case the attachment map $(L, \partial) \to (L \,\widehat{\amalg}\, \overline{\mathbb{L}}_{sT}, \partial)$ identifies up to quasi-isomorphism with the projection $(L, \partial) \to (L', \partial)$ and is clearly surjective in homology. On the other hand by Theorem 16.2, the homotopy Lie algebra of the homotopy fiber of the attaching map $\langle L \rangle \to \langle L \,\widehat{\amalg}\, \overline{\mathbb{L}}_{sT} \rangle$ is the profree Lie algebra $\overline{\mathbb{L}}_{(T \widehat{\otimes} \overline{UL'})}$.

In this section we show that inert maps have always a similar form.

We begin with a definition. Let U and V be closed subspaces of an enriched Lie algebra L. We denote by (U', V') a new copy of (U, V) and denote by $F(U, V)$ the closure of the subset of $L \,\widehat{\amalg}\, L'$ generated by the iterated brackets $[x_1, [\ldots, [x_{n-1}, x_n] \ldots]$ with $x_i \in U \subset L$ or $x_i \in V' \subset L'$. We finally denote by $\pi : L \,\widehat{\amalg}\, L' \to L$ the morphism that identifies L' with L.

Definition The *(generalized) free product* of U and V, denoted $U \,\widehat{\amalg}\, V$, is the image of $F(U, V)$ along π.

Now, let $T \subset L$ be a closed subspace, and let I be the closed ideal generated by T in L. We denote by $\sigma : (L/I) \to L$ a linear section of the projection $L \to L/I$.

Definition The ideal I generated by T is *freely generated* if:

(1) The natural map $\overline{\mathbb{L}}_T \to L$ is an injection.

(2) The injection

$$\overline{\mathbb{L}}_T \,\widehat{\amalg}\, \sigma(L/I) \to L$$

is an isomorphism of graded vector spaces.

Definition Let I be a closed ideal in an enriched Lie algebra. The *I-graded Lie algebra of L* is the Lie algebra

$$g_I(L) = \prod_{n \geq 0} g_I(L)_n, \quad \text{with } g_I(L)_n = I^{(n)}/I^{(n+1)},$$

with the graded Lie bracket

$$[I^{(r)}/I^{(r+1)}, I^{(s)}/I^{(s+1)}] \subset I^{(r+s)}/I^{(r+s+1)}.$$

Remark that $g_I(L)_0 = L/I$ is a sub-Lie algebra of $g_I(L)$.

Theorem 16.3 *Let T be a closed subset in an enriched Lie algebra L; then the following conditions are equivalent:*

(i) The attaching map $\overline{\mathbb{L}}_T \to L$ is inert.
(ii) The ideal I generated by T is freely generated.
(iii) $g_I(L) \cong \overline{\mathbb{L}}_T \,\widehat{\amalg}\, L/I$ as a graded Lie algebra.

Proof (i) $\implies$ (ii). Since the attachment is inert, the ideal I generated by T is profree and $I = \overline{\mathbb{L}}_{T \widehat{\otimes} \overline{UL/I}}$ (Theorem 16.2). In particular each element of L is in the image of $\overline{\mathbb{L}}_T \,\widehat{\amalg}\, \sigma(L/I)$. This implies (ii).

(ii) $\implies$ (iii). By construction $L = I \oplus \sigma(L/I)$. In particular, $g_I(L) = \overline{\mathbb{L}}_T \,\widehat{\amalg}\, L/I$.

(iii) $\implies$ (i). The injection $\overline{\mathbb{L}}_T \to g_I(L)$ is clearly inert, and therefore I is profree and $I/I^2 = T \widehat{\otimes} \overline{UL/I}$. This implies (i) by Theorem 16.2. $\qquad\square$

Corollary *A subspace $S \subset T$ of an inert subspace is also inert.*

Proof Write $T = S \oplus R$. Since T is inert, the injection $\overline{\mathbb{L}}_T \,\widehat{\amalg}\, \sigma(L/T) \to L$ is an isomorphism of graded vector spaces. Write $\overline{\mathbb{L}}_T = \overline{\mathbb{L}}_S \,\widehat{\amalg}\, \overline{\mathbb{L}}_R$ to get an isomorphism $\overline{\mathbb{L}}_S \,\widehat{\amalg}\, \sigma(L/S)$, where $L/S = \overline{\mathbb{L}}_R \,\widehat{\amalg}\, L/T$. $\qquad\square$

16.8 Pairs of dgl Attaching Maps

Suppose that $\gamma : (\overline{\mathbb{L}}_T, 0) \to (L, \partial)$ is a dgl attaching map and that

$$T = T(1) \oplus T(2)$$

decomposes T as a direct sum of closed subspaces. This yields dgl attaching maps

$$\gamma(1) : (\overline{\mathbb{L}}_{T(1)}, 0) \to (L, \partial) \qquad \text{and } \gamma(2) : (\overline{\mathbb{L}}_{T(2)}, 0) \to (L \,\widehat{\amalg}\, \overline{\mathbb{L}}_{sT(1)}, \partial).$$

Proposition 16.6 *With the hypotheses and notation above,*

$$\gamma \text{ is inert} \iff \gamma(1) \text{ and } \gamma(2) \text{ are both inert.}$$

Proof Denote by

$$j(1) : L \to L \,\widehat{\amalg}\, \overline{\mathbb{L}}_{sT(1)} \qquad \text{and } j(2) : L \,\widehat{\amalg}\, \overline{\mathbb{L}}_{sT(1)} \to L \,\widehat{\amalg}\, \overline{\mathbb{L}}_{sT(1)} \,\widehat{\amalg}\, \overline{\mathbb{L}}_{sT(2)} = L \,\widehat{\amalg}\, \overline{\mathbb{L}}_{sT}$$

the inclusions corresponding to $\gamma(1)$ and $\gamma(2)$. By definition, $j : L \to L \,\widehat{\amalg}\, \overline{\mathbb{L}}_{s(T)}$ is given by

$$j = j(2) \circ j(1).$$

In particular, if $\gamma(1)$ and $\gamma(2)$ are inert, then $H(j) = H(j(2)) \circ H(j(1))$ is surjective and by definition γ inert.

In the reverse direction, suppose γ is inert. Since $H(j)$ is surjective, $H(j(2))$ is also surjective, and therefore $\gamma(2)$ is inert. To show that $\gamma(1)$ is inert, we may assume that the differential in L is zero (Proposition 16.3). In this case, since γ is inert, Theorem 16.3 asserts that $\ker H(j)$ is freely generated by $\gamma(\overline{\mathbb{L}}_T)$. In other words, we have an isomorphism

$$(\overline{\mathbb{L}}_{T(1)} \,\widehat{\amalg}\, \overline{\mathbb{L}}_{T(2)}) \,\widehat{\amalg}\, \sigma(L/I) \xrightarrow{\;\cong\;} L.$$

On the other hand, denote by I_2 the ideal generated by $T(2)$ in $H(L \,\widehat{\amalg}\, \overline{\mathbb{L}}_{sT(1)})$; then

$$H(L \,\widehat{\amalg}\, \overline{\mathbb{L}}_{sT(1)})/(I_2) = L/I.$$

Since γ_2 is inert, we have a linear bijection

$$H(L \,\widehat{\amalg}\, \overline{\mathbb{L}}_{sT(1)}) \cong \overline{\mathbb{L}}_{T(2)} \,\widehat{\amalg}\, \sigma(L/I).$$

Since we can choose $\overline{\mathbb{L}}_{T(2)}\,\widehat{\amalg}\,\sigma(L/I)$ as the image of a section of $H(L\,\widehat{\amalg}\,\overline{\mathbb{L}}_{sT(1)}) \to L/I$, the linear bijection

$$\overline{\mathbb{L}}_{T(1)}\,\widehat{\amalg}\,\left(\overline{\mathbb{L}}_{T(2)}\,\widehat{\amalg}\,\sigma(L/I)\right) \xrightarrow{\;\cong\;} L$$

shows that $\gamma(1)$ is inert. $\qquad\square$

Example In [18, Theorem 2.2], S. Chenery proves the following statement: $f : \Sigma A \to X \vee Y$ is a continuous map whose composition with the quotient $q : X \vee Y \to X$ is inert; then f is also inert. Our next proposition is the corresponding statement for enriched dgl's and is a direct consequence of Proposition 16.6.

Proposition 16.7 *Let* $f : \overline{\mathbb{L}}_Z \to L_1\,\widehat{\amalg}\,L_2$ *be a morphism of enriched graded Lie algebras, and let* $q : L_1\,\widehat{\amalg}\,L_2 \to L_1$ *be the projection on the first factor. If* qf *is an inert map, then* f *is an inert map.*

Proof We consider first the case $L_2 = \overline{\mathbb{L}}_R$ and apply Proposition 16.6 with $T = Z \oplus R$, $\gamma(1)$ is the identity on R, and $\gamma(2) = qf$. Then $\gamma(1)$ and $\gamma(2)$ being inert, γ is inert.

In the general case, let $\overline{\mathbb{L}}_R \to L_2$ be a profree dgl model. Then the result follows from the commutative diagram

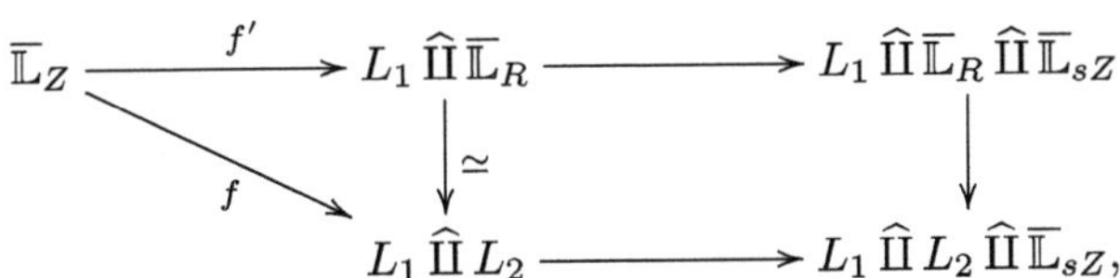

where f' is a lift of f. $\qquad\square$

16.9 Free Products

Proposition 16.8 *Suppose* $Q \subset L$ *and* $Q' \subset L'$ *are inert subspaces of enriched Lie algebras. Then*

$$Q \oplus Q' \subset L\,\widehat{\amalg}\,L'$$

is inert.

In particular, $Q \subset L\,\widehat{\amalg}\,L'$ *is inert.*

Proof By hypothesis we have dgl's $(L\,\widehat{\amalg}\,\overline{\mathbb{L}}_N, \partial)$ and $(H(L'\,\widehat{\amalg}\,\overline{\mathbb{L}}_{N'}, \partial')$ with $\partial : N \xrightarrow{\cong} Q$ and $\partial' : N' \xrightarrow{\cong} Q'$ and

$$H(L) \to H'L\,\widehat{\amalg}\,\overline{\mathbb{L}}_N) \quad \text{and} \quad H(L') \to H(L'\,\widehat{\amalg}\,\overline{\mathbb{L}}_{N'})$$

both surjective.

Since $H(L\,\widehat{\amalg}\,L') \cong H(L)\,\widehat{\amalg}\,H(L')$, this is a direct consequence of the following commutative diagram in which ψ is surjective:

$$
\begin{array}{ccc}
H(L\,\widehat{\amalg}\,L') & \longrightarrow & H(L\,\widehat{\amalg}\,L'\,\widehat{\amalg}\,\overline{\mathbb{L}}_N\,\widehat{\amalg}\,\overline{\mathbb{L}}_{N'}) \\
\Big\downarrow{\scriptstyle\cong} & & \Big\downarrow{\scriptstyle\cong} \\
H(L)\,\widehat{\amalg}\,H(L') & \xrightarrow{\ \psi\ } & H(L\,\widehat{\amalg}\,\overline{\mathbb{L}}_N,\partial)\,\widehat{\amalg}\,H(L'\,\widehat{\amalg}\,\overline{\mathbb{L}}_{N'},\partial').
\end{array}
$$

$\square$

16.10   Inertness of Elements in a Profree Lie Algebra

Theorem 16.4 *Suppose $\overline{\mathbb{L}}_T$ is a profree Lie algebra in which $T = T_{even}$. Then any nonzero morphism*

$$\gamma : \overline{\mathbb{L}}_{\mathbb{Q}x} \to \overline{\mathbb{L}}_T$$

is an inert attaching map.

Proof By Proposition 16.3 we can suppose that the differential is zero in $\overline{\mathbb{L}}_T$. We accomplish the proof in three steps.

Step 1. *Reduction to the case $T = T_0$*

We suppose the result is true when $W = W^1$, and we prove it in general.

Since $(\wedge W, d)$ is quadratic and $W = W^{odd}$, we can associate with $\wedge W$ a second gradation given by the length of the words. Then putting S in gradation 1, the attaching map $\rho : \wedge W \to \mathbb{Q} \oplus S$ is a morphism of bigraded cdga's. The injection $A \to \wedge W$ has therefore also a bigraded minimal model $\wedge V \to \wedge W$. Forgetting the original gradation, we are in the situation W is in gradation zero, which implies that $\wedge V \to \wedge W$ is an injective map.

Step 2. *Reduction to the case $T = T_0$ and $\dim T < \infty$*

Here we suppose that the result is true when $\dim T < \infty$ and prove it in general.

By definition T is equipped with a family of surjections $\rho_\alpha : T \to T_\alpha$ onto finite dimensional vector spaces and

$$\overline{\mathbb{L}}_T = \varprojlim_{\alpha} \overline{\mathbb{L}}_{T_\alpha}.$$

Therefore, if $x \in \overline{\mathbb{L}}_T$ is nonzero, then some $x_\alpha = \overline{\mathbb{L}}_{\rho_\alpha} x$ is nonzero. On the other hand $T = \ker \rho_\alpha \oplus Q$ with $\rho_\alpha : Q \xrightarrow{\cong} T_\alpha$. It is immediate that $\overline{\mathbb{L}}_{\ker \rho_\alpha} \to \overline{\mathbb{L}}_T$ is inert. Since $\overline{\mathbb{L}}_{Qx} \to \overline{\mathbb{L}}_{T_\alpha}$ is inert,

$$\overline{\mathbb{L}}_{Qx \oplus \ker \rho_\alpha} \to \overline{\mathbb{L}}_T$$

is inert. Therefore by Proposition 16.6, $\rho : \overline{\mathbb{L}}_{Qx} \to \overline{\mathbb{L}}_T$ is inert.

Step 3. *Proof when $T = T_0$ and dim $T < \infty$*

By definition, γ is inert if and only if

$$\overline{\mathbb{L}}_T \to H(\overline{\mathbb{L}}_T \,\widehat{\amalg}\, \mathbb{L}_{Qy})$$

is surjective where the differential in $\overline{\mathbb{L}}_T \,\widehat{\amalg}\, \mathbb{L}_{Qy}$ is defined by $\partial y = x$. Now fix a basis $x_1, \ldots, x_n$ of T, and observe that $\overline{\mathbb{L}}_T \,\widehat{\amalg}\, \mathbb{L}_{Qy}$ is the completion of the free graded Lie algebra

$$L = \mathbb{L}_{(x_1, \ldots, x_n, y)}.$$

Denote by $L(k)$ the subspace of L of iterated Lie brackets of length k in the x_i and y. Thus an element in $(\overline{\mathbb{L}}_T \,\widehat{\amalg}\, \mathbb{L}_{Qy})_p$ is an infinite sum of the form $z = \sum z_k$, with each $z_k \in L(k)_p$. In particular, for some $r \geq 1$,

$$\partial y = \sum_{k \geq r} a_k$$

with $a_k \in L(k)_0$ and $a_r \neq 0$.

Now consider the dgl $(L, \bar{\delta})$ with $\bar{\delta} y = a_r$. For this dgl we can modify the degrees in L, as follows: Simply assign deg 2 to each x_i and degree $2r + 1$ to y. It follows from ([41], Theorem 3.12) that with this new gradation $\overline{\mathbb{L}}_T \to H(L, \bar{\delta})$ is surjective. Therefore $H(L, \partial) = \overline{\mathbb{L}}_T/(x)$ is a graded Lie algebra for the graduation given by the lower central series. This then remains true with the original gradation: $H_{\geq 1}(L, \bar{\delta}) = 0$.

Next, let ω be a cycle of degree $p \geq 1$ in $\overline{\mathbb{L}}_T \,\widehat{\amalg}\, \mathbb{L}_{Qy}$, and write

$$\omega = \sum_{k \geq m} \omega_k, \quad \text{with } \omega_k \in (L(k)_p,$$

and $\omega_m \neq 0$. Then $\bar{\delta}\omega_m \in L(m + r - 1)_{p-1}$ and

$$\partial\omega - \bar{\delta}\omega_m \in L(\geq m + r)_{>p-1}.$$

It follows that $\bar{\delta}\omega_m = 0$ and so ω_m is a $\bar{\delta}$-boundary:

$$\omega_m = \overline{\delta}\Phi_{m-r+1}, \quad \text{some } \Phi_{m-r+1} \in L(m-r+1)_{p+1}.$$

But this in turn implies that

$$\omega - \partial\Phi_{m-r+1} \in L(\geq m+1).$$

Iterating this process yields a sequence $\Phi_{k-r+1} \in L(k-r+1)_{p+1}$ with

$$\partial\left(\sum_k \Phi_{k-r+1}\right) = \omega.$$

$\square$

Example Let $\omega_1, \ldots, \omega_n$ be a series of elements in a profree Lie algebra $\overline{\mathbb{L}}_T$. This induces a morphism of enriched dgl's

$$f : \overline{\mathbb{L}}(y_1, \ldots, y_n) \to \overline{\mathbb{L}}_T$$

given by $f(y_i) = \omega_i$.

Now we suppose that T contains closed subsets $T(1) \subset \cdots \subset T(n) = T$ such that $\omega_i \in \overline{\mathbb{L}}_{T(i)}$ and $\omega_i \neq 0$ in the quotient dgl $\overline{\mathbb{L}}_{T(i)/T(i-1)}$. Then the morphism f is inert. This follows directly from Theorem 17.1 with an iterated application of Proposition 16.6.

For instance, the sequence $[x_1, x_2], [x_2, x_3], \ldots, [x_{n-1}, x_n]$ is inert in $\overline{\mathbb{L}}(x_1, \ldots, x_n)$.

16.11   A Connected CW Complex, Y, for Which $\pi_*(Y_{\mathbb{Q}}) = \pi_1(Y_{\mathbb{Q}})$, $\dim H_1(Y; \mathbb{Q}) = 3$, $\dim H_2(Y; \mathbb{Q}) = \infty$, and $H_p(Y; \mathbb{Q}) = 0$ for $p > 2$

Denote by $X = S_a^1 \vee S_b^1 \vee S_c^1$ a wedge of three circles. Then $\pi_1(X)$ is the free group, G, on generators a, b, c defined by the inclusions of the circles. Thus the profree dgl $(L_X, 0) = (\overline{\mathbb{L}}(x, y, z), 0)$ is a profree dgl model for X where $x, y, z = \log_G a, \log_G b, \log_G c$ (Proposition 9.3(i)).

Now use $\overline{ad}(-)$ to denote commutation with elements in G

$$\overline{ad}(u)(v) = (u, v) = uvu^{-1}v^{-1}.$$

Then a sequential cell attachment

$$X \longrightarrow Y = \varinjlim_r (X \cup_g \vee_{n=1}^r D_n^2)$$

is defined by $\pi_1(g)[S_n^1] = \overline{ad}^n(a)\overline{ad}^n(b)(c)$. A profree dgl representative for this attachment is then provided by the sequence

$$(L_X, 0) \to \varinjlim_r (L_X \,\widehat{\amalg}\, \mathbb{L}_{sT(r)}, \partial)$$

where $T(r) = \oplus_{n=1}^r \mathbb{Q}z_n$ and $\partial s z_n = \log_G(\overline{ad}^n(a)\overline{ad}^n(b)(c))$. Since $\partial : sT(r) \to L_X^2$, it follows from Proposition 14.1 that

$$\dim H_p(X \cup_g \vee_{n=1}^r D_n^2; \mathbb{Q}) = \begin{cases} 3 \text{ if } p = 1, \\ r \text{ if } p = 2, \\ 0 \text{ if } p > 2. \end{cases}$$

It follows from the obvious direct limit argument that

$$\dim H_p(Y; \mathbb{Q}) = \begin{cases} 3 & \text{if } p = 1, \\ \infty & \text{if } p = 2, \\ 0 & \text{if } p > 2. \end{cases}$$

Denote by $ad(-)$ the conjugation in $L_X : ad_x(y) = [x, y]$. Then, for some element $\gamma_n \in L_X^{>2n+1}$, we have

$$\log_G(\overline{ad}^n(a)\overline{ad}^n(b)(c)) = ad^n(x)ad^n(y)(z) + \gamma_n.$$

Now consider the dgl morphism

$$(L_X, 0) \to (L_X \,\widehat{\amalg}\, \mathbb{L}_{sT(r)}, \partial'),$$

where

$$\partial' s z_n = ad^n(x)ad^n(y)(z).$$

Write $L = L_X \,\widehat{\amalg}\, \mathbb{L}_{sT(r)}$, and suppose we have proved that $H_q(L, \partial') = 0$ for $q > 1$. We introduce an extra gradation in L by putting x, y, and z in degree 1 and z_n in degree $2n + 1$. We denote $L_{(n)}$ the component of L in the new degree n. Then $\partial' : L_{(n)} \to L_{(n)}$, and $(\partial - \partial')(L_{(n)}) \subset \oplus_{q>n} L_{(q)}$.

Suppose ω is a ∂-cycle in $L^{\geq 2}$, and write $\omega = \sum_{q \geq q_0} \omega_q$ with $\omega_q \in L_{(q)}$. Then ω_{q_0} is a ∂'-cycle and so a ∂'-boundary. We write $\omega_{q_0} = \partial'(\beta_0)$ with $\beta_0 \in L_{(q_0)}$. Then denote $\omega' = \omega - \partial(\beta_0)$. We have $\omega' = \sum_{q>q_0} \omega'_q$, ω'_{q_0+1} is a ∂'-cycle, so a boundary, and there is $\beta_1 \in L_{(q_0+1)}$ with $\partial'\beta_1 = \omega'_{q_0+1}$. We continue and construct in this way a sequence of elements $\beta_q \in L_{(q_0+q)}$. The series $\sum \beta_q$ is a well-defined element in L and

$$\partial\left(\sum \beta_q\right) = \omega.$$

This shows that $H_{\geq 2}(L, \partial) = 0$.

Now we prove that $H_{\geq 2}(L, \partial') = 0$. Let $(\wedge W, d)$ be the quadratic Sullivan model of $L_X = \overline{\mathbb{L}}(x, y, z)$. Proposition 6.4 equips $(\wedge W, d)$ with a second gradation $W = \oplus_{n \geq 0} W(n)$ satisfying

$$W(0) = W \cap \ker d \quad \text{and } d : W(n) \to (\wedge^2 W)(n - 1).$$

Then denote by w_x, w_y, and w_z the basis of $W(0)$ which is the dual basis for sx, sy, and sz.

Next set $\alpha_n = ad^n(x)ad^n(y)(z)$. Then α_n determines a cdga morphism $\rho_n : \wedge W \to \mathbb{Q} \oplus \mathbb{Q}a_n$ and satisfying

$$\rho_n(W) \neq 0 \quad \text{and } \rho_n(W(k)) = 0 \quad \text{if } k \neq 2n.$$

Set $S(r) = \oplus_{n=1}^{r} \mathbb{Q}a_n$. Then

$$\rho(r) = \oplus_{n=1}^{r} \rho_n : \wedge W \to \mathbb{Q} \oplus S(r)$$

is well defined and surjective.

Now we introduce a third gradation in $\wedge W$ by assigning $W(k)$ the new degree $2k + 3$. Then d increases this new degree by 1, so that $\wedge W$ is identified with a quadratic Sullivan algebra $\wedge Z$ in which $W(k) = Z^{2k+3}$ and $Z = Z^{\geq 3}$. In particular, a_n has new degree $2(2n) + 3$. Thus $\rho(r)$ with this new grading becomes a surjective morphism

$$\rho'(r) : \wedge Z \to \mathbb{Q} \oplus S'(r).$$

Thus $\wedge Z \xrightarrow{\cong} \mathbb{Q} \oplus (Z \cap \ker d)$, and $Z \cap \ker d = W(0)$ but with new degree 3. It follows that the homotopy Lie algebra of $\wedge Z$ is the free graded Lie algebra on the elements x, y, z but regarded as having degree 3. We now show that the sequence α_n, regarded as elements in $L_{\wedge Z}$, is inert.

Recall from [5, Theorem 1.4] that the sequence α_n is inert in $L_{\wedge T}$ if and only if it is inert in its universal enveloping algebra. Moreover following [5, Theorem 2.1], if for some ordering of a, b, c the family w_i of the terms of maximum height of the α_i satisfies the following two conditions, then the family α_i is inert. The conditions are:

(1) No w_i is a submonomial of any other w_j.
(2) No w_i "overlaps with" any other w_j, i.e., if $w_i = uv$ and $w_j = bw$, then $v = 1$.

In our situation let $x > y > z$. Then the maximum height in α_n is $x^n y^n z$, and this family satisfies the two conditions above.

It follows that the sequence α_n is inert in $\wedge Z$. This implies that we can suppose the induced morphism $\lambda' : \wedge V' \to \wedge Z$ satisfies $\lambda'(V') \subset Z$.

Now return to the original gradings. Since $\wedge V \cong \wedge Z$ and $\lambda = \lambda'$, it follows that $\lambda : V \to W$ is injective, and by Theorem 16.2 this shows that the sequence α_n is inert in L_X. In particular, for each r,

$$L_X \to H(L_X \,\widehat{\amalg}\, \overline{\mathbb{L}}_{sT(r)}, \partial') \quad \text{and} \quad L_X \to H(L_X \,\widehat{\amalg}\, \overline{\mathbb{L}}_{sT(r)}, \partial)$$

are surjective.

Since $\dim T(r) < \infty$,

$$X_\mathbb{Q} \to (X \cup_g \vee_{n=1}^{r} D_n^2)_\mathbb{Q}$$

is the spatial realization of $L_X \to H(L_X \,\widehat{\amalg}\, \overline{\mathbb{L}}_{sT(r)})$. This identifies the surjection $L_X \to H(L_X \,\widehat{\amalg}\, \overline{\mathbb{L}}_{sT(r)})$ with

$$\pi_*(X_\mathbb{Q}) \to \pi_*((X \cup_g \vee_{n=1}^{r} D_n^2)_\mathbb{Q}).$$

Since $L_X = (L_X)_0$, $\pi_*(X_\mathbb{Q}) = \pi_1(X_\mathbb{Q})$ and so $\pi_{\geq 2}(X \cup_g \vee_{n=1}^{r} D_n^2)_\mathbb{Q} = 0$. Passage to direct limits shows that $\pi_*(Y_\mathbb{Q}) = \pi_1(Y_\mathbb{Q})$. $\qquad\square$

Applications in Topology 17

In this chapter we give a list of topological applications of the results developed in the previous chapters. In particular we extend the inert property of the last cell of a Poincaré duality complex to the general connected case. We also give a global description of the rational homotopy Lie algebra of a 2-cone.

17.1 Characterization of Rationally Inert Topological Attaching Maps

Let $f : \vee_\alpha S^{n_\alpha} \to X$ be a continuous map of connected spaces and $i_X : X \to Y = X \cup_f (\vee_\alpha D^{n_\alpha+1})$ the associated attachment map. We denote by F the homotopy fiber of $(i_X)_\mathbb{Q} = X_\mathbb{Q} \to Y_\mathbb{Q}$, by T the image of $\oplus_\alpha \pi_{n_\alpha}(S^{n_\alpha}) \otimes \mathbb{Q}$ in $L = \pi_*(X_\mathbb{Q})$, and by I the closed ideal generated by T in L.

Proposition 17.1 *The following conditions are equivalent:*

(i) f is rationally inert.
(ii) F has the rational homotopy type of a wedge of spheres.
(iii) I is a profree Lie algebra, and the right adjoint representation of L/I in $I/I^{(2)}$ induces an isomorphism of $\overline{UL/I}$-modules

$$T \,\widehat{\otimes}\, \overline{UL/I} \xrightarrow{\;\cong\;} I/I^{(2)}.$$

(iv) The graded Lie algebra $g_I(L) = \prod_n I^n/I^{n+1}$ satisfies

$$g_I(L) \cong \mathbb{L}_T \,\widehat{U}\, L/I.$$

© The Author(s), under exclusive license to Springer Nature Switzerland AG 2026
Y. Félix, S. Halperin, *Lie Models for Spaces*, Frontiers in Mathematics,
https://doi.org/10.1007/978-3-032-15357-9_17

Proof (i) $\Leftrightarrow$ (ii) is Theorem 16.2. The equivalence (i)–(iii) is a corollary of Theorem 14.1, and the equivalence (i)–(iv) follows from Theorem 16.4. $\square$

17.2 Whitehead's Problem

A famous unsolved problem of JHC Whitehead [83] asks: Is a subcomplex of an aspherical two-dimensional CW complex aspherical? As observed by Anick [4] it is sufficient to consider the case that the complex and the subcomplex share the same 1-skeleton and base point. The problem then reduces to the question: If X is a finite two-dimensional connected CW complex and $X \cup \left(\vee_{k=1}^{p} D_k^2 \right)$ is aspherical, is X aspherical?

In [4] Anick provides a positive answer to an analogous question for simply connected rational spaces. Here we have a positive answer for Sullivan rationalizations of connected spaces.

Theorem 17.1 *If X is a connected CW complex and $(X \cup \vee_{k=1}^{p} D_k^2)_{\mathbb{Q}}$ is aspherical, then $X_{\mathbb{Q}}$ is aspherical.*

Proof The obvious induction reduces the statement to the case $p = 1$. Then, by hypothesis $\pi_*((X \cup_f D^2)_{\mathbb{Q}}) \cong V^{\vee}$ with $V = V^1$. Let $\varphi : (\wedge V, d) \to (\wedge W, d)$ be a Sullivan representative for the inclusion $i : X \to X \cup_f D^2$. Since $H^1(X \cup_f D^2) \to H^1(X)$ is injective, it follows that φ is injective and so $\wedge W$ decomposes as $\wedge V \otimes \wedge Z$, with $Z = Z^{\geq 1}$. In particular f is rationally inert. Moreover, it follows that

$$H^{\geq 1}(\wedge Z, \overline{d}) \cong \mathbb{Q}b \otimes \wedge U,$$

where $\deg b = 1$ and $\wedge V \otimes \wedge U$ is the acyclic closure of $\wedge V$. Since $V = V^1$, $U = U^0$ and $H^{\geq 1}(\wedge Z, \overline{d}) = H^1(\wedge Z, \overline{d})$. This in turn implies $Z = Z^1$, and $X_{\mathbb{Q}}$ is aspherical. $\square$

17.3 Poincaré Duality Complexes

For recall *a Poincaré duality algebra of dimension n* is a pair (A, ω) in which A is a connected graded commutative algebra and $\omega : A^n \to \mathbb{Q}$ is a linear map such that the induced bilinear forms

$$A^k \otimes A^{n-k} \to \mathbb{Q}, \quad a \otimes b \mapsto \omega(ab)$$

are nondegenerate.

We say a CW complex $Y = X \cup_f D^{n+1}$ is a *rational Poincaré duality complex* if $H(Y)$ is a Poincaré duality algebra and the top class is in the image of $H(Y, X)$. In this case it follows that $H^{\leq n}(Y) \xrightarrow{\cong} H(X)$.

Theorem 17.2 *If $Y = X \cup_f D^{n+1}$ is a rational Poincaré duality complex and the algebra $H(Y)$ requires at least two generators, then f is rationally inert.*

This theorem was first established for simply connected spaces in [41, Theorem 5.1]. For non-simply connected spaces this is a particular case of the following proposition.

Proposition 17.2 *Let X be an n-dimensional connected space and $Y = X \cup_f e^{n+1}$ such that:*

(i) $H^{n+1}(M) \cong \mathbb{Q}\omega$.
(ii) There are elements $v \in H^1(M)$ and $w \in H^n(M)$ such that $wv = \omega$.

Then f is an inert attachment.

Proof Recall that f is rationally inert if

$$\pi_*(i_\mathbb{Q}) : \pi_*(X_\mathbb{Q}) \to \pi_*(X \cup_f D^{n+1})_\mathbb{Q}$$

is surjective. Now let $\lambda : \wedge V \to \wedge W$ be a Sullivan representative of $i : X \to X \cup_f D^{n+1}$, and then decompose $\wedge W$ as a Λ-extension

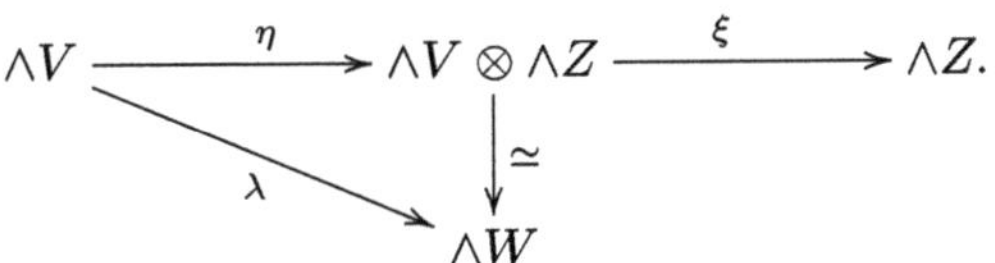

Then according to Theorem 16.1, the map f is rationally inert if and only if the homotopy Lie algebra, L_Z, is profree, i.e., (Theorem 6.1)

$$\wedge Z \simeq (\mathbb{Q} \oplus S, 0)$$

with $S \cdot S = 0$.

Now let $\wedge V \otimes \wedge U$ be the acyclic closure of $\wedge V$. Then we have

$$\wedge W \otimes \wedge U := \wedge W \otimes_{\wedge V} (\wedge V \otimes \wedge U) \simeq \wedge Z.$$

Therefore we have only to prove that

$$\wedge W \otimes \wedge U \simeq (\mathbb{Q} \oplus S, 0)$$

with $S \cdot S = 0$.

For this, we first establish the following notation. Let P be a direct summand in $(\wedge V)^{n+1}$ of $(\wedge V)^{n+1} \cap \ker d$. Then division by P and by $(\wedge V)^{>n+1}$ defines a surjective quasi-isomorphism $\wedge V \xrightarrow{\simeq} A$, and

$$A^{n+1} = A^{n+1} \cap \operatorname{Im} d \oplus \mathbb{Q}\omega,$$

where ω is a cycle representing the top cohomology class of Y.

In the same way, let P' be a direct summand in $(\wedge W)^n$ of $(\wedge W)^n \cap \ker d$. Then division by P' and $(\wedge W)^{>n}$ defines a surjective quasi-isomorphism $\wedge W \xrightarrow{\simeq} B$. Moreover, by construction φ factorizes in a commutative diagram

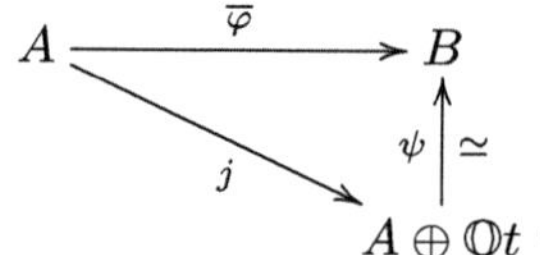

in which $\overline{\varphi}\omega = 0$.

Now consider the commutative cdga diagram

$$
\begin{array}{ccc}
A & \xrightarrow{\overline{\varphi}} & B \\
 & \searrow{\scriptstyle j} & \uparrow{\scriptstyle \psi}\ \simeq \\
 & & A \oplus \mathbb{Q}t
\end{array}
$$

in which $dt = \omega$, $t \cdot A^{\geq 1} = 0$, $t^2 = 0$, and $\psi t = 0$. Here ψ is a quasi-isomorphism, and j is a cdga representative of φ. This provides a cdga quasi-isomorphism

$$\wedge W \otimes \wedge U \simeq (A \oplus \mathbb{Q}t) \otimes_A (A \otimes_{\wedge V} (\wedge V \otimes \wedge U)) = (A \oplus \mathbb{Q}t) \otimes \wedge U.$$

Thus from the short exact sequence of chain complexes

$$0 \to A \otimes \wedge U \to (A \oplus \mathbb{Q}t) \otimes \wedge U \to \mathbb{Q}t \otimes \wedge U \to 0,$$

we deduce a linear isomorphism

$$H^{\geq 1}(\wedge W \otimes \wedge U) \cong H^{\geq 1}((A \oplus \mathbb{Q}t) \otimes \wedge U) \xrightarrow{\cong} \mathbb{Q}t \otimes \wedge U$$

of graded vector spaces. It remains to show that we can lift a basis of $\mathbb{Q}t \otimes \wedge U$ to cycles $\Phi_i \in (A \oplus \mathbb{Q}t) \otimes \wedge U$ such that $\Phi_i \cdot \Phi_j = 0$.

Before going further we first eliminate two special cases. First, if $V^1 = 0$, the argument of [41, §5] shows that $(A \oplus \mathbb{Q}t) \otimes \wedge U$ is rationally wedge-like, and so f is rationally inert. (Note that in [41] it is assumed that X is simply connected; however the proof of this assertion relies only on the fact that $V^1 = 0$.) Second, if $n = 1$, then $X \simeq_{\mathbb{Q}} S_1^1 \vee \cdots \vee S_{2q}^1$, and so Y is rationally equivalent to an oriented Riemann surface. In this case Theorem 17.3 is established in [29]. $\qquad\square$

Thus we may now assume that $n \geq 2$ and that A^1 contains a nonzero cycle v. Since $H(A)$ is a Poincaré duality algebra, there is a cycle $w \in A^n$ such that $wv = \omega$. The first step for the proof is then given by the next Lemma 17.1.

Lemma 17.1 *With the hypotheses and notation above, $A^{n+1} \otimes \wedge U \subset d(A^n \otimes \wedge U)$.*

Proof Choose $\overline{v} \in U^0$ so that $d\overline{v} = v$. Since $\wedge V$ is a minimal Sullivan algebra, V is the union of an increasing sequence of subspaces $V(0) \subset \cdots \subset V(q) \subset \ldots$ in which $V(0) = \mathbb{Q}v$ and $d : V(q+1) \to \wedge V(q)$. It follows that U is the union of an increasing sequence of subspaces $U(0) \subset \cdots \subset U(q) \subset \ldots$ in which $U(0) = \mathbb{Q}\overline{v}$ and

$$d : U(q+1) \to A^{\geq 1} \otimes \wedge U(q).$$

We show by induction on q that

$$A^{n+1} \otimes \wedge U(q) \subset d(A^n \otimes \wedge U(q))$$

First note that any $z \in A^{n+1}$ has the form $z = dy + \lambda wv$, some $\lambda \in \mathbb{Q}$. Thus

$$z \otimes 1 = d(y \otimes 1) + (-1)^n d(\lambda w \otimes \overline{v}) \in d(A^n \otimes \wedge U(0)).$$

Then for $r \geq 1$,

$$z \otimes \overline{v}^r = d(y \otimes \overline{v}^r + \frac{(-1)^n \lambda}{r+1} w \otimes \overline{v}^{r+1}) - (-1)^n ryv \otimes \overline{v}^{r-1}.$$

It follows by induction on r that $A^{n+1} \otimes \wedge U(0) \subset d(A^n \otimes \wedge U(0))$.

Now fix a direct summand, T, of $U(q)$ in $U(q+1)$, and assume by induction that for some s,

$$A^{n+1} \otimes \wedge U(q) \otimes \wedge^{\leq s} T \subset d(A^n \otimes \wedge U(q) \otimes \wedge^{\leq s} T).$$

Then write $\Phi \in A^{n+1} \otimes \wedge U(q) \otimes \wedge^{s+1} T$ as $\Phi = \sum \Phi_i \otimes \Psi_i$ with $\Phi_i \in A^{n+1} \otimes \wedge U(q)$ and $\Psi_i \in \wedge^{s+1} T$. By the hypothesis $\Phi_i = d\Omega_i$ with $\Omega_i \in A^n \otimes \wedge U(q)$. Therefore

$$\sum \Phi_i \otimes \Psi_i = d\left(\sum \Omega_i \otimes \Psi_i\right) - (-1)^n \sum \Omega_i \wedge d\Psi_i.$$

The first term is in $d(A^n \otimes \wedge U(q) \otimes \wedge^{s+1} T)$. On the other hand, $d\Psi_i \in A^{\geq 1} \otimes \wedge U(q) \wedge^{\leq s} T$ and so the second term is in $A^{n+1} \otimes \wedge U(q) \otimes \wedge^{\leq s} T$. By hypothesis, the second term is contained in $d(A^n \otimes \wedge U(q) \otimes \wedge^{\leq s} T)$. This closes the induction. $\qquad\square$

Now, let $\Phi \in \wedge U$. Then

$$t - (-1)^n w\overline{v} \in (A \oplus \mathbb{Q}t) \otimes \wedge U$$

is a cycle, and

$$d((t - (-1)^n w\overline{v})\Phi) = -w\overline{v}\, d\Phi \in A^{n+1} \otimes \wedge U.$$

By Lemma 17.1, $w\overline{v}\, d\Phi = d\Psi$ for some $\Psi \in A^n \otimes \wedge U$. Thus $(t - (-1)^n)w\overline{v})\Phi + \Psi$ is a cycle projecting to $t \otimes \Phi$ in $\mathbb{Q}t \otimes \wedge U$. Thus such cycles map to a basis of $\mathbb{Q}t \otimes \wedge U$. But because $n \geq 2$, $2n > n + 1$ and so the product of any two of those cycles is zero. This shows that f is rationally inert. $\qquad\square$

17.4 Inertness of the Adjunction of a Cell to a Wedge of Circles

For simplicity we say that $[f] \in \pi_1(X)$ is rationally inert if $f : S^1 \to X$ is rationally inert.

Theorem 17.3 *If X is a wedge of at least two circles, then any nonzero $[f] \in \pi_1(X)$ is rationally inert, or, equivalently, $(X \cup_f D^2)_{\mathbb{Q}}$ is aspherical.*

Proof Since the homotopy Lie algebra, L_X, is profree and concentrated in degree zero, by Theorem 16.4, every nonzero element in L_X is inert: The morphism $\pi_1(X_{\mathbb{Q}}) \to \pi_*((X \cup_f D^2)_{\mathbb{Q}})$ is surjective. Since $\pi_*(X_{\mathbb{Q}})$ is concentrated in degree 1, $(X \cup_f D^2)_{\mathbb{Q}}$ is aspherical. $\qquad\square$

17.5 The Homotopy Lie Algebra of a 2-Cone

A 2-*cone* is a space $Y = (\vee_j S^{n_j}) \cup_g (\vee_i D^{k_i+1})$ constructed from an attaching map $g : \vee_i S^{k_i} \to X = \vee_j S^{n_j}$. We denote by

$$f : X \to Y = X \cup_g (\vee_i D^{k_i+1})$$

the induced inclusion. We assume that $H^{\geq 1}(g) = 0$ and that $\pi_*(g_{\mathbb{Q}})$ is injective, and we denote by $\varphi : (\overline{\mathbb{L}}_S, 0) \to (\overline{\mathbb{L}}_{S \oplus sT}, \partial)$ a minimal dgl representative of f.

Proposition 17.3

(i) *Denote by E the image of L_X in L_Y. Then $E = L_X/I$, where I is the kernel of $H_*(\varphi)$. Moreover I is a profree Lie algebra generated by the $\overline{UE}$-module generated by $\partial(sT)$.*

(ii) *There is a retraction $r : L_Y \to E$ whose kernel is a profree Lie algebra K, i.e., there is a short exact sequence*

$$0 \to K \to L_Y \to E \to 0.$$

In particular, the attaching map g is rationally inert if and only if $K = 0$, and otherwise L_Y contains a profree Lie algebra.

Proof First of all remark that we have a commutative diagram

$$
\begin{array}{ccc}
L_X & \xrightarrow{\ L_f\ } & L_Y \\
\cong \big\uparrow & & \cong \big\uparrow \\
\overline{\mathbb{L}}_S & \xrightarrow{\ H(\varphi)\ } & H(\overline{\mathbb{L}}_{S \oplus sT}, \partial) \text{'}
\end{array}
$$

(i) We equip $\overline{\mathbb{L}}_{S \oplus sT}$ with an extra lower gradation with S in gradation 0 and sT in gradation 1. The differential ∂ decreases the gradation by 1, and it follows that I is the closed ideal generated by $\partial(sT)$. Since I is a closed ideal in the profree Lie algebra $\overline{\mathbb{L}}_S$, I is profree and generated by $\partial(sT)$.

(ii) Let $(\wedge W, d)$ be a quadratic model of X, and let $\rho : \wedge W \to \mathbb{Q} \oplus S$, $S \cdot S = 0$, the map associated with g. Then we form the cdga's $(\wedge W \oplus s^{-1}S, D)$ and $(\wedge W \oplus S \oplus s^{-1}S, D)$, with $Dw = dw + s^{-1}\rho(w)$ and $Dx = -s^{-1}x$ for $x \in S$. The following commutative diagram

$$
\begin{array}{ccccccccc}
0 & \longrightarrow & (\wedge W \oplus s^{-1}S, D) & \longrightarrow & (\wedge W \oplus S \oplus s^{-1}S, D) & \longrightarrow & S & \longrightarrow & 0 \\
& & \cong \big\uparrow & & \cong \big\uparrow \theta & & \big\| & & \\
0 & \longrightarrow & \ker \rho & \longrightarrow & \wedge W & \longrightarrow & S & \longrightarrow & 0 \text{,}
\end{array}
$$

in which $\theta(w) = w + \rho(w)$, shows that $(\wedge W \oplus s^{-1}S, D)$ is a cdga model for L_Y.

We equip $\wedge W \oplus s^{-1}S$ with a new gradation $\wedge W \oplus s^{-1}S = \oplus_{k \geq 0}(\wedge W \oplus s^{-1}S)^{(k)}$ with $W = W^{(1)}$ and $s^{-1}S = (s^{-1}S)^{(2)}$. Then D increases this gradation by 1. Therefore the cohomology of $\wedge W \oplus s^{-1}S$ inherits a new gradation

$$H^*(\wedge W \oplus s^{-1}S, D) = \oplus_{p \geq 0} H^{(p)}.$$

The short exact sequence

$$0 \to (s^{-1}S, 0) \to (\wedge W \oplus s^{-1}S, D) \to (\wedge W, d) \to 0$$

shows that $H^{(p)} = 0$ for $p > 2$. On the other hand, clearly, $H^{(0)} = \mathbb{Q}$, and, since X is a wedge of spheres, $H^{(1)} = H^{\geq 1}(X)$.

Now, let $\varphi : (\wedge V, d) \xrightarrow{\simeq} (\wedge W \oplus s^{-1}S, D)$ be a Sullivan minimal model. We can suppose $(\wedge V, d)$ equipped with a new gradation preserved by φ. In particular, $(\wedge V^{(1)}, d)$ is a sub cdga, and its homotopy Lie algebra is isomorphic to E. The injection $(\wedge V^{(1)}, d) \to (\wedge V, d)$ induces the surjection $L_Y \to E$.

Denote by $(\wedge V^{(1)} \otimes \wedge U, d)$ the acyclic closure of $(\wedge V^{(1)}, d)$ equipped with the new gradation given by $U = U^{(0)}$. Then let $\wedge Z$ be the Sullivan minimal model of $(\wedge V \otimes \wedge U, d)$. Then $Z = \oplus_{p \geq 2} V^{(p)}$, and since $H(\wedge V \otimes \wedge U, d) = \mathbb{Q} \oplus H^{(2)}(\wedge V \otimes \wedge U, d)$, the cdga $(\wedge Z, d)$ is quasi-isomorphic to $(\mathbb{Q} \oplus (Z \cap \ker d), 0)$. Thus L_Z is profree, and we get the required extension

$$0 \to L_Z \to L_Y \to E \to 0.$$

In the notation of the proposition, $K = L_Z$ $\square$

17.6   The Homotopy Fiber of the Rationalization of a Cell Adjonction

Let $f : S^n \to X$ be a continuous map into a connected space with $n \geq 1$. We denote by $Y = X \cup_f e^{n+1}$ the result of the adjonction of a cell along f and by $\varphi : (\overline{\mathbb{L}}_V, d) \to (\overline{\mathbb{L}}_{V \oplus \mathbb{Q}x}, d)$ an enriched dgl model for the injection $\iota : X \to Y$. By construction $\omega = d(x)$ is a cycle representative of the image of $[f]$ along the rationalization $X \to X_{\mathbb{Q}}$.

From φ we form the commutative diagram

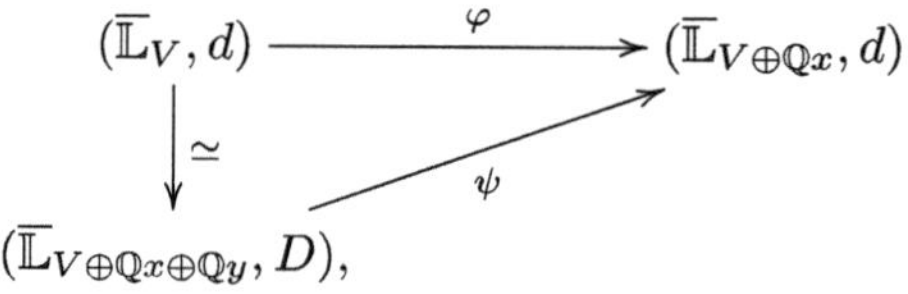

where $Dy = 0$, $Dx = \omega - y$, and $Dv = dv$ for $v \in V$.

Proposition 17.4 *With the above notations:*

(i) The space $\langle \mathrm{Ker}\, \psi \rangle$ has the homotopy type of the homotopy fiber F of $\iota_{\mathbb{Q}} : X_{\mathbb{Q}} \to Y_{\mathbb{Q}}$.

(ii) $\operatorname{Ker}\psi \cong (\overline{\mathbb{L}}_W, D)$, with $W = \widehat{T}(V \oplus \mathbb{Q}x) \otimes y$. In the image of $\operatorname{Ker}\psi$ in $(\overline{\mathbb{L}}_{V \oplus \mathbb{Q}x \oplus \mathbb{Q}y}, D)$, the space $\widehat{T}(V \oplus \mathbb{Q}x)$ acts on y by the adjoint action.

Proof

(i) The Sullivan model of a short exact sequence of enriched dgl's is a Λ-extension, and by [33, Proposition 17.9], its topological realization is a fibration.

(ii) If we forget the differential, then y is an inert element in $\overline{\mathbb{L}}_{V \oplus \mathbb{Q}x \oplus \mathbb{Q}y}$. The result follows then directly from Theorem 16.3.

$\square$

Corollary *Let $\omega \in L_n$ and $a \in L_{2r}$, $2r < n$, be non-homologically trivial cycles in an enriched dgl L, and consider the dgl extension $j : (L, d) \to (L', d) = (L \,\widehat{\amalg}\, \overline{\mathbb{L}}_{(x)}, d)$, $dx = \omega$. Then:*

(i) *If ω is inert, then $[a, \omega] \neq 0$ in $H(L)$.*

(ii) *If $[a, \omega] = 0$, then ω is not inert, and $H(L')$ contains a profree Lie algebra on two generators F such that $\operatorname{Im} j \cap F = 0$.*

Proof

(i) Since ω is inert, the ideal I generated by ω is $\overline{\mathbb{L}}_{\overline{UH(L')} \otimes \omega}$, and $[a]$ is a nonzero element in $H(L')$.

(ii) Since $[a, \omega]$ is a boundary in L, there is an element $u \in L$ such that $du = [a, \omega]$. Then $\alpha = [a, x] - u$ is a cycle in L', and $\beta = [a, [a, x]] - [a, u]$ is also a cycle.

We denote by R the quotient enriched dgl

$$R = L/(L_{\geq n} \oplus d(L_n)).$$

Denote then by q the natural projection $L' \to R' = (R \,\widehat{\amalg}\, \overline{\mathbb{L}}_{(x)}, d')$. The images by q of the cycles α and β are, respectively, $[a, x]$ and $[a, [a, x]]$. On the other hand $d'x = 0$. Therefore x becomes an inert element, and by Theorem 16.3, the ideal generated by x is $\overline{\mathbb{L}}_{\overline{UR} \otimes x}$. In particular $[a, x]$ and $[a, [a, x]]$ generate a nontrivial profree Lie algebra in $H(R')$. This implies that the Lie algebra generated by α and β is also a profree Lie algebra in $H(L')$.

$\square$

17.7   The Fiber of a Successive Addition of Cells

Proposition 17.5 *Let Y be a connected space obtained by a successive addition of cells e_α to an n-dimensional space X. We suppose that for each α, $\dim e_\alpha \in [n+1, 2n+1]$ and that the induced map $\psi : \pi_*(X_\mathbb{Q}) \to \pi_*(Y_\mathbb{Q})$ is surjective. Then $\ker \psi$ is a profree Lie algebra.*

Remark that $\ker\psi$ is the homotopy Lie algebra of the homotopy fiber of the injection $X_\mathbb{Q} \to Y_\mathbb{Q}$.

Proof Let $\varphi : (\overline{\mathbb{L}}_V, d) \to (\overline{\mathbb{L}}_{V\oplus W}, d)$ be a dgl profree model for the injection $X \hookrightarrow Y$. Then φ factorizes as

$$(\overline{\mathbb{L}}_V, d) \xrightarrow{\ j\ } (\overline{\mathbb{L}}_{V\oplus W\oplus sW}, D) \xrightarrow{\ q\ } (\overline{\mathbb{L}}_{V\oplus W}, d),$$

where D is defined degree by degree by

$$\begin{cases} Dv = dv \\ Dw = dw + sw \\ Dsw = -Ddw \end{cases}$$

Then j is a quasi-isomorphism, and q is surjective. It follows that

$$\ker q = H(\overline{\mathbb{L}}_{\overline{U\overline{\mathbb{L}}_{V\oplus W}}\,\widehat{\otimes}\, sW},\, D).$$

Since the degree of an element of sW is in the interval $[n, 2n]$, $D(sW) \subset \overline{U\overline{\mathbb{L}}_{V\oplus W}}\,\widehat{\otimes}\, sW$. Therefore, since $H_*(q)$ is surjective,

$$\ker q \cong \overline{\mathbb{L}}_{H(Z)}$$

where Z is the complex

$$(\overline{U\overline{\mathbb{L}}_{V\oplus W}}\,\widehat{\otimes}\, sW,\, D).$$

$\square$

References

1. Allday, C., Puppe, V.: Cohomological Methods in Transformation Groups. Cambridge Studies in Advanced Mathematics, vol. 32. Cambridge University Press, Cambridge (1993)
2. Amman, M.: The Omnibus conjecture - disproved. Preprint 2020, arXiv 2011.01827
3. Anick, D.: Non-commutative algebras and their Hilbert series. J. Algebra **78**, 120–140 (1982)
4. Anick, D.: A rational homotopy analogue of Whitehead's problem. In: Algebra, Algebraic Topology and its Applications. Lecture Notes in Mathematics, vol. 1183, pp. 28–31. Springer, Berlin (1986)
5. Anick, D.: Inert sets and the Lie algebra associated to a group. J. Algebra **111**, 154–165 (1987)
6. Barratt, M.G., Milnor, J.: An example of anomalous singular homology. Proc. Amer. Math. Soc. **13**, 293–297 (1962)
7. Berglund, A., Madsen, I.: Rational homotopy theory of automorphisms of manifolds. Acta Mathematica **224**, 67–185 (2020)
8. Berglund, A., Saleh, B.: A dg Lie model for relative homotopy automorphisms. Homol. Homotopy Appl. **22**, 105–121 (2020)
9. Bourbaki, N.: Eléments de Mathématique, Théorie des Ensembles, tome 3, Hermann (1970)
10. Bousfield, A.K.: Localization and periodicity in unstable homotopy theory. J. Amer. Math. Soc. **7**, 831–874 (1994)
11. Bousfield, A.K., Gugenheim, V.K.: On PL de Rham Theory and Rational Homotopy Type. Memoirs of the American Mathematical Society, vol. 179. American Mathematical Society, Providence (1976)
12. Bousfield, A.K., Kan, D.M.: Homotopy Limits, Completions and Localizations. Lecture Notes in Mathematics, vol. 304. Springer, Berlin (1972)
13. Brown, K.: Cohomology of Groups. Graduate Texts in Mathematics, vol. 87. Springer, Berlin (1982)
14. Bubenik, P.: Free and semi-inert cell attachments. Trans. Amer. Math. Soc. **357**, 4533–4553 (2005)
15. Buijs, U., Félix, Y., Murillo, A., Tanré, D.: Lie Models in Topology. Progress in Mathematics, vol. 335. Birkhauser, Basel (2020)
16. Calaque, D., Campos, R., Nuiten, J.: Lie algebroids and curved Lie algebras. Preprint (2021)
17. Chen, K.-T.: Extension of C^∞ function algebra by integrals and Malcev completion of π_1. Adv. Math. **23**, 181–210 (1977)
18. Chenery, S.: Loop space decompositions of connected sums and applications to the Vigué-Poirrier conjecture, PhD Thesis (2023)
19. Chuang, J., Lazarev, A., Mannan, W.H.: Cocommutative coalgebras: homotopy theory and Koszul duality. Homol. Homotopy Appl. **18**, 303–336 (2016)

Y. Félix, S. Halperin, *Lie Models for Spaces*, Frontiers in Mathematics,
https://doi.org/10.1007/978-3-032-15357-9

20. Cirici, J.: A short course on the interactions of rational homotopy theory and complex (algebraic) geometry, preprint Barcelona (2020)
21. Cornea, O., Lupton, G., Oprea, J., Tanré, D.: Lustenik-Schnirelmann Category. Mathematical Surveys and Monographs, vol. 103. American Mathematical Society, Providence (2001)
22. Dror, Farjoun, E.: Cellular Spaces, Null Spaces, and Homotopy Localization. Lecture Notes in Mathematics, vol. 1622. Springer, Berlin (1995)
23. Dwyer, W.G., Spalinski, J.: Homotopy theories and model categories. In: Handbook of Algebraic Topology, pp. 73–126. North-Holland, Amsterdam (1995)
24. Dyer, E., Vasquez, A.: Some small aspherical spaces. J. Austr. Math. Soc. **16**, 332–352 (1973)
25. Félix, Y., Halperin, S.: Malcev completions, LS category and depth. Bol. Soc. Math. Mex. **23**, 267–288 (2017)
26. Félix, Y., Halperin, S.: Rational homotopy theory via Sullivan spaces: a survey. In: ICCM Notices of the International Congress of Chinese Mathematicians, vol. 5, pp. 14–36 (2017)
27. Félix, Y., Halperin, S.: The Sdepth of a homotopy Lie algebra. Geometry Topol. Math. Phys. J. **1**, 4–26 (2018)
28. Félix, Y., Halperin, S.: The depth and LS category of a topological space. Math. Scand. **123**, 220–238 (2018)
29. Félix, Y., Halperin, S.: The depth of a Riemann surface and of a right-angled Artin group. J. Homotopy Related Struct. **15**, 223–248 (2020)
30. Félix, Y., Tanré, D.: Rational homotopy of the polyhedral product functor. Proc. Amer. Math. Soc. **137**, 891–898 (2008)
31. Félix, Y., Thomas, J.-C.: Module d'holonomie d'une fibration. Bull. Math. Soc. France **113**, 255–258 (1985)
32. Félix, Y., Thomas, J.-C.: Le Tor différentiel d'une fibration non nilpotente. J. Pure Appl. Algebra **38**, 217–233 (1985)
33. Félix, Y., Halperin, S., Thomas, J.-C.: Rational Homotopy Theory. Graduate Texts in Mathematics, vol. 201. Springer, Berlin (2001)
34. Félix, Y., Halperin, S., Thomas, J.C.: The ranks of the homotopy groups of a space of finite LS category. Expos. Math. **25**, 67–76 (2007)
35. Félix, Y., Oprea, J., Tanré, D.: Algebraic Models in Geometry. Oxford Graduate Texts in Mathematics, vol. 17. Oxford University Press, Oxford (2008)
36. Félix, Y., Halperin, S., Thomas, J.-C.: Rational Homotopy Theory II. World Scientific, Singapore (2015)
37. Getzler, E.: Lie theory for nilpotent L_∞-algebras. Ann. Math. **170**, 271–301 (2009)
38. Goerss, P.: The homology of homotopy inverse limits. J. Pure Appl. Algebra **111**, 83–122 (1996)
39. Hain, R.: The de Rham homotopy theory of complex algebraic varieties I. K-Theory **1**, 271–324 (1987)
40. Hall, M.: A basis for free Lie rings and higher commutators in free groups. Proc. Amer. Math. Soc. **1**, 575–581 (1950)
41. Halperin, S., Lemaire, J.M.: Suites inertes dans les algèbres de Lie graduées. Math. Scand **61**, 39–67 (1987)
42. Halperin, S., Lemaire, J.-M.: The fibre of a cell attachment. Proc. Roy. Soc. Edinburgh **38**, 295–311 (1995)
43. Halperin, S., Stasheff, J.D.: Obstructions to homotopy equivalences. Adv. Math. **32**, 233–279 (1979)
44. Halperin, S., Gomez-Tato, A., Tanré, D.: Rational homotopy theory for non-simply connected spaces. Trans. Amer. Math. Soc. **352**, 1493–1525 (2000)
45. Hamilton, A.: A Poincaré-Birkhoff-Witt theorem for profinite pronilpotent Lie algebras. J. Lie Theory **29**, 611–618 (2019)

46. Hess, K.: Rational Homotopy Theory : A Brief Introduction. Preprint (2006)
47. Hess, K., Lemaire, J.-M.: Nice and lazy cell attachments. J. Pure Appl. Algebra **112**, 29–39 (1996)
48. Hilton, P., Mislin, G., Roitberg, J.: Localization of Nilpotent Groups and Spaces. North-Holland Mathematics Studies, vol. 15. North-Holand, Amsterdam (1975)
49. Hinich, V.: Descent of Deligne groupoids. Inst. Math. Res. Not. **1997**, 223–239 (1997)
50. Hinich, V.: DG coalgebras as formal stacks. J. Pure Appl. Algebra **162**, 209–250 (2001)
51. Humphreys, J.E.: Linear Algebraic Groups. Graduate Texts in Mathematics, vol. 31. Springer, Berlin (1975)
52. Holstein, J.: Rational Homotopy Theory. Master Course 2020/21. University of Hamburg, Hamburg
53. Ivanov, S.: An overview of rationalization theories of non-simply connected spaces and non-nilpotent groups. Acta Mathematica Sinica **38**, 1705–1721 (2022)
54. Ivanov, S., Mikhailov, R.: A finite $\mathbb{Q}$-bad space. Geometry Topol. **23**, 1237–1249 (2019)
55. Jech, Th.: Set Theory, the Third Millennium Edition, Revised and Expanded. Springer Monographs in Mathematics. Springer, Berlin (2003)
56. Lawrence, R., Sullivan, D.: A formula for topology/deformations and its signifiance. Fund. Math. **225**, 229–242 (2014)
57. Lazard, M.: Sur les groupes nilpotents et les anneaux de Lie. Ann. Sc. Ec. Norm. Sup. **71**, 101–190 (1959)
58. Lazarev, A., Markl, M.: Disconnected rational homotopy theory. Adv. Math. **283**, 303–361 (2015)
59. Lemaire, J.-M.: A finite complex whose rational homology is not finitely generated. In: H Spaces. Lecture Notes in Mathematics, vol. 196, pp. 114–120 (1971)
60. Lemaire, J.-M.: "Autopsie d'un meurtre" dans l'homologie d'une algèbre de chaînes. Ann. Scient. Ec. Norm. Sup. **11**, 93–100 (1978)
61. Libgober, A.: Families of singular algebraic varieties that are rationally elliptic spaces, preprint (2025). arXiv 2501.17970v1
62. Lyndon, R.C.: Cohomology theory of groups with a single defining relation. Ann. Math. **52**, 650–656 (1950)
63. Majewski, M.: Rational Homotopical Models and Uniqueness. Memoirs of the American Mathematical Society, vol. 682. American Mathematical Society, Providence (2000)
64. May, J.P.: Simplicial Objects in Algebraic Topology. University of Chicago Press, Chicago (1967)
65. May, J.P., Ponto, K.: More Concise Algebraic Topology: Localization, Completion and Model Categories. Chicago University Press, Chicago (2011)
66. Milivojevic, A., Stelzig, J., Zoller, L.: Formality is preserved under domination, preprint (2023). arXiv 2306.12364v1
67. Milnor, J., Moore, J.C.: On the structure of Hopf algebras. Ann. Math. **81**, 211–264 (1965)
68. Neisendorfer, J.: A quick trip through localization. In: Alpine Perspective on Algebraic Topology. Contemporary Mathematics, vol. 504. American Mathematical Society, Providence (2009)
69. Papadima, S., Suciu, A.: Algebraic invariants for right-angled Artin groups. Math. Ann. **334**, 533–555 (2006)
70. Putman, A.: One-relator groups. Preprint (2018)
71. Pridham, J.P.: Pro-algebraic homotopy types. Proc. London Math. Soc. **97**, 273–338 (2008)
72. Quillen, D.: Rational homotopy theory. Ann. Math. **90**, 205–295 (1969)
73. Quillen, D.: Homotopical Algebra. Lecture Notes in Mathematics, vol. 43. Springer, Berlin (1967)

74. Serre, J.P.: Lie Algebras and Lie Groups. Benjamin, Chuo City (1965)
75. Shirshov, A.I.: Subalgebras of free Lie algebras. Mat. Sbornik **33**, 441–452 (1975)
76. Solovay, R.: $2^{\aleph_0}$ can be anything it ought to be. In: Symposium on the Theory of Models, pg 435. North-Holland Publish, Amsterdam (1965)
77. Sullivan, D.: Infinitesimal computations in topology. Publ. IHES **47**, 269–331 (1977)
78. Tanré, D.: Homotopie Rationnelle: Modèles de Chen, Quillen, Sullivan. Lecture Notes in Mathematics, vol. 1025. Springer, Berlin (1983)
79. Theriault, S.: Homotopy Fibrations with a Section After Looping. To Appear Memoirs of the American Mathematical Society. American Mathematical Society, Providence (2024)
80. Theriault, S.: Top cell attachment for a Poincaré duality complex. Preprint (2024)
81. Voronov, A.: Rational Homotopy Theory. Encyclopedia of Mathematical Physics, vol. 4, pp. 24–38, 2nd edn. Academic, Cambridge (2025)
82. Watkiss, C.: Unpublished pager (1976)
83. Whitehead, J.H.C.: On adding relations to homotopy groups. Ann. Math. **42**, 409–428 (1941)
84. Zhou, J.: Cohomology of minimal Sullivan algebras of non-finite type and their realizations. ArXiv 2024

Index

Y. Félix, S. Halperin, *Lie Models for Spaces*, Frontiers in Mathematics,
https://doi.org/10.1007/978-3-032-15357-9